AF892965

324

Topics in Current Chemistry

Editorial Board:
K.N. Houk • C.A. Hunter • M.J. Krische • J.-M. Lehn
S.V. Ley • M. Olivucci • J. Thiem • M. Venturi • P. Vogel
C.-H. Wong • H. Wong • H. Yamamoto

Topics in Current Chemistry

Recently Published and Forthcoming Volumes

Activity-Based Protein Profiling
Volume Editor: Stephan A. Sieber
Vol. 324, 2012

Beauty in Chemistry
Volume Editor: Luigi Fabbrizzi
Vol. 323, 2012

Constitutional Dynamic Chemistry
Volume Editor: Mihail Barboiu
Vol. 322, 2012

EPR Spectroscopy
Volume Editors: Malte Drescher, Gunnar Jeschke
Vol. 321, 2012

Radicals in Synthesis III
Volume Editors: Markus R. Heinrich, Andreas Gansäuer
Vol. 320, 2012

Chemistry of Nanocontainers
Volume Editors: Markus Albrecht, F. Ekkehardt Hahn
Vol. 319, 2012

Liquid Crystals: Materials Design and Self-Assembly
Volume Editor: Carsten Tschierske
Vol. 318, 2012

Fragment-Based Drug Discovery and X-Ray Crystallography
Volume Editors: Thomas G. Davies, Marko Hyvönen
Vol. 317, 2012

Novel Sampling Approaches in Higher Dimensional NMR
Volume Editors: Martin Billeter, Vladislav Orekhov
Vol. 316, 2012

Advanced X-Ray Crystallography
Volume Editor: Kari Rissanen
Vol. 315, 2012

Pyrethroids: From Chrysanthemum to Modern Industrial Insecticide
Volume Editors: Noritada Matsuo, Tatsuya Mori
Vol. 314, 2012

Unimolecular and Supramolecular Electronics II
Volume Editor: Robert M. Metzger
Vol. 313, 2012

Unimolecular and Supramolecular Electronics I
Volume Editor: Robert M. Metzger
Vol. 312, 2012

Bismuth-Mediated Organic Reactions
Volume Editor: Thierry Ollevier
Vol. 311, 2012

Peptide-Based Materials
Volume Editor: Timothy Deming
Vol. 310, 2012

Alkaloid Synthesis
Volume Editor: Hans-Joachim Knölker
Vol. 309, 2012

Fluorous Chemistry
Volume Editor: István T. Horváth
Vol. 308, 2012

Multiscale Molecular Methods in Applied Chemistry
Volume Editors: Barbara Kirchner, Jadran Vrabec
Vol. 307, 2012

Solid State NMR
Volume Editor: Jerry C. C. Chan
Vol. 306, 2012

Activity-Based Protein Profiling

Volume Editor: Stephan A. Sieber

With Contributions by

R. Breinbauer · R. Bischoff · P.P. Geurink · W.P. Heal · Y.-G. Kim ·
J. Krysiak · G.A. van der Marel · M.B. Nodwell · H.S. Overkleeft ·
L.M. Prely · A. Saghatelian · S.A. Sieber · E.W. Tate

Springer

Editor
Prof. Dr. Stephan A. Sieber
Department of Organic Chemistry II
Technical University of Munich
Garching
Germany

ISSN 0340-1022 e-ISSN 1436-5049
ISBN 978-3-642-28377-2 e-ISBN 978-3-642-28378-9
DOI 10.1007/978-3-642-28378-9
Springer Heidelberg Dordrecht London New York

Library of Congress Control Number: 2012932058

© Springer-Verlag Berlin Heidelberg 2012
This work is subject to copyright. All rights are reserved, whether the whole or part of the material is concerned, specifically the rights of translation, reprinting, reuse of illustrations, recitation, broadcasting, reproduction on microfilm or in any other way, and storage in data banks. Duplication of this publication or parts thereof is permitted only under the provisions of the German Copyright Law of September 9, 1965, in its current version, and permission for use must always be obtained from Springer. Violations are liable to prosecution under the German Copyright Law.
The use of general descriptive names, registered names, trademarks, etc. in this publication does not imply, even in the absence of a specific statement, that such names are exempt from the relevant protective laws and regulations and therefore free for general use.

Printed on acid-free paper

Springer is part of Springer Science+Business Media (www.springer.com)

Volume Editor

Prof. Dr. Stephan A. Sieber

Department of Organic Chemistry II
Technical University of Munich
Garching
Germany

Editorial Board

Prof. Dr. Kendall N. Houk

University of California
Department of Chemistry and Biochemistry
405 Hilgard Avenue
Los Angeles, CA 90024-1589, USA
houk@chem.ucla.edu

Prof. Dr. Christopher A. Hunter

Department of Chemistry
University of Sheffield
Sheffield S3 7HF, United Kingdom
c.hunter@sheffield.ac.uk

Prof. Michael J. Krische

University of Texas at Austin
Chemistry & Biochemistry Department
1 University Station A5300
Austin TX, 78712-0165, USA
mkrische@mail.utexas.edu

Prof. Dr. Jean-Marie Lehn

ISIS
8, allée Gaspard Monge
BP 70028
67083 Strasbourg Cedex, France
lehn@isis.u-strasbg.fr

Prof. Dr. Steven V. Ley

University Chemical Laboratory
Lensfield Road
Cambridge CB2 1EW
Great Britain
Svl1000@cus.cam.ac.uk

Prof. Dr. Massimo Olivucci

Università di Siena
Dipartimento di Chimica
Via A De Gasperi 2
53100 Siena, Italy
olivucci@unisi.it

Prof. Dr. Joachim Thiem

Institut für Organische Chemie
Universität Hamburg
Martin-Luther-King-Platz 6
20146 Hamburg, Germany
thiem@chemie.uni-hamburg.de

Prof. Dr. Margherita Venturi

Dipartimento di Chimica
Università di Bologna
via Selmi 2
40126 Bologna, Italy
margherita.venturi@unibo.it

Prof. Dr. Pierre Vogel

Laboratory of Glycochemistry
and Asymmetric Synthesis
EPFL – Ecole polytechnique féderale
de Lausanne
EPFL SB ISIC LGSA
BCH 5307 (Bat.BCH)
1015 Lausanne, Switzerland
pierre.vogel@epfl.ch

Prof. Dr. Chi-Huey Wong

Professor of Chemistry, Scripps Research
Institute
President of Academia Sinica
Academia Sinica
128 Academia Road
Section 2, Nankang
Taipei 115
Taiwan
chwong@gate.sinica.edu.tw

Prof. Dr. Henry Wong

The Chinese University of Hong Kong
University Science Centre
Department of Chemistry
Shatin, New Territories
hncwong@cuhk.edu.hk

Prof. Dr. Hisashi Yamamoto

Arthur Holly Compton Distinguished
Professor
Department of Chemistry
The University of Chicago
5735 South Ellis Avenue
Chicago, IL 60637
773-702-5059
USA
yamamoto@uchicago.edu

Topics in Current Chemistry Also Available Electronically

Topics in Current Chemistry is included in Springer's eBook package *Chemistry and Materials Science*. If a library does not opt for the whole package the book series may be bought on a subscription basis. Also, all back volumes are available electronically.

For all customers with a print standing order we offer free access to the electronic volumes of the series published in the current year.

If you do not have access, you can still view the table of contents of each volume and the abstract of each article by going to the SpringerLink homepage, clicking on "Chemistry and Materials Science," under Subject Collection, then "Book Series," under Content Type and finally by selecting *Topics in Current Chemistry*.

You will find information about the

- Editorial Board
- Aims and Scope
- Instructions for Authors
- Sample Contribution

at springer.com using the search function by typing in *Topics in Current Chemistry*.

Color figures are published in full color in the electronic version on SpringerLink.

Aims and Scope

The series *Topics in Current Chemistry* presents critical reviews of the present and future trends in modern chemical research. The scope includes all areas of chemical science, including the interfaces with related disciplines such as biology, medicine, and materials science.

The objective of each thematic volume is to give the non-specialist reader, whether at the university or in industry, a comprehensive overview of an area where new insights of interest to a larger scientific audience are emerging.

Thus each review within the volume critically surveys one aspect of that topic and places it within the context of the volume as a whole. The most significant developments of the last 5–10 years are presented, using selected examples to illustrate the principles discussed. A description of the laboratory procedures involved is often useful to the reader. The coverage is not exhaustive in data, but rather conceptual, concentrating on the methodological thinking that will allow the non-specialist reader to understand the information presented.

Discussion of possible future research directions in the area is welcome.

Review articles for the individual volumes are invited by the volume editors.

In references *Topics in Current Chemistry* is abbreviated *Top Curr Chem* and is cited as a journal.

Impact Factor 2010: 2.067; Section "Chemistry, Multidisciplinary": Rank 44 of 144

Preface

An overarching challenge in the field of proteomics is the assignment of the molecular, cellular and physiological functions to the proteome - the entire compliment of proteins encoded by an organism's genome. As protein functions are governed by a vast array of post-translational modifications, methods to directly investigate native protein activity against a background of high biological complexity have been developed. At the forefront of these techniques is the use of small molecule probes that display selective interaction with active proteins, labelling them and allowing a fractionation of the proteome based on activity. This method, named Activity Based Protein Profiling (ABPP), is emerging as a mature discipline in the field of proteomics, and allows for the identification, classification and biochemical characterization of low abundance proteins that would otherwise remain undetected.

In this special volume of Topics of Current Chemistry, we have assembled review articles from various experts in the field of ABPP. Topics covered include the use of natural products and photoaffinity labels in ABPP, the use of metabolomic approaches for functional protein characterization and the use of ABPP in mapping microbial pathogenesis.

The first article by Nodwell and Sieber provides a general introduction, surveying analytical platforms for ABPP, enzymatic classes addressable by ABPP probes, and biological applications of ABPP.

The term "activity-based protein profiling" implies mechanism-based probe/target reactivity. Photoaffinity labelling approaches represent a complementary technique to mechanism based APBB probes. The use of these photoreactive "affinity-based protein profiling" probes in proteomic studies are reviewed by Overkleeft et al.

Natural products, by virtue of their co-evolution with protein systems, possess integrated biointeractvity. The unique properties of natural products render them ideal scaffolds for activity-based protein probes. Krysiak and Breinbauer cover the use of natural products in ABPP.

With the development of antibiotic resistant bacterial strains such as methicillin-resistant *Staphylococcus aureus* (MRSA), infectious diseases have become again a life-threatening problem, raising concerns of a return to a "pre-antibiotic" era. A key approach to arresting the spread of resistant bacteria is the understanding of pathogenesis. Heal and Tate focus on the use of ABPP to discovering and mapping pathways of pathogenesis in bacteria.

Many enzymes that are identified via ABPP are uncharacterized and many of them play crucial roles in pathogenesis. In order to understand the function of these enzymes in more detail their substrates have to be investigated via metabolomic procedures. Kim and Saghatelian highlight this important technology.

We, the volume editors, hope that this book will prove to be a helpful reference for students as well as the advanced researcher. We have attempted to provide a background to ABPP as well as review some of the more intriguing applications of this emerging technology. At the time of writing, this field is far from being comprehensively explored, and novel innovations and applications of ABPP will undoubtedly provide insight into protein function for many years to come.

Stephan Sieber and Matthew Nodwell

Contents

Top Curr Chem (2012) 324: 1–42
DOI: 10.1007/128_2011_302
© Springer-Verlag Berlin Heidelberg 2011
Published online: 13 December 2011

ABPP Methodology: Introduction and Overview

Matthew B. Nodwell and Stephan A. Sieber

Abstract Activity-based protein profiling (ABPP) is emerging as a mature method for chemically interrogating the proteome of a cell. This chapter serves to introduce the reader to ABPP by providing overviews of the general principles of the technique, analytical methods used in ABPP, the classes of enzymes that can be specifically addressed by ABPP probes, and biological applications of ABPP.

Keywords Activity-based protein profiling · Analytical chemistry · Chemical biology · Proteomics

Contents

M.B. Nodwell and S.A. Sieber (✉)
Department Chemie, Centre for Integrated Protein Science CIPSM, Institute of Advanced Studies, Technische Universität München, Lichtenbergstrasse 4, 85747 Garching, Germany
e-mail: stephan.sieber@tum.de

Abbreviations

(TOP)-ABPP	Tandem orthogonal proteolysis ABPP
3-oxo-C12-HSL	3-Oxo-dodecanoyl homoserine lactone
ABPP	Activity-based protein profiling
ADP	Adenosine diphosphate
AOMK	Acyloxymethyl ketone
ASPP	Active site peptide profiling
ATP	Adenosine triphosphate
BODIPY	Boron-dipyrromethane
CE	Capillary electrophoresis
FDA	Food and Drug Administration
FP	Fluorophosphonate
HDAC	Histone deacetylase
LC-MS	Liquid chromatography–mass spectrometry
LIF	Laser-induced fluorescence
ML	Mixed lineage
MP	Metalloprotease
MS	Mass spectrometry
MudPIT	Multidimensional protein identification technology
PAD4	Protein arginine deiminase 4
PI3K	Phosphoinositide 3-kinase
PKMT	Protein lysine methyltransferase
PNA	Peptide nucleic acid
RA	Rheumatoid arthritis
SAHA	Suberoylanilide hydroxamic acid
SAM	*S*-Adenosyl-L-methionine
SDS	Sodium dodecylsulfate
SDS-PAGE	Sodium dodecylsulfate polyacrylamide gel electrophoresis
TEV	Tobacco etch virus

1 Introduction

Novel therapeutic targets are urgently needed for the treatment of infections and several devastating diseases such as cancer and diabetes. As the approval of novel drugs becomes more and more sophisticated due to the high measures of safety, the in-depth characterization of a drug–target pair as well as of all putative off-targets is of paramount importance. Moreover, the mining for novel, so far unexplored, targets is an additional challenge that requires novel methodologies in the fields of chemistry, biology, and analytical sciences. As the technological toolbox for target discovery has rapidly developed over the past decade, several complex biological questions can be now be addressed using standard procedures. However, a major breakthrough was achieved by the completion of several genome projects.

Genome sequencing projects have provided an enormous amount of information on gene identities in prokaryotic and eukaryotic organisms. In the context of proteomics, this accumulation of knowledge has led to the challenge of assigning the molecular, cellular, and physiological functions for the full compliment of proteins encoded by these genomes. Analysis of genomic data itself can offer some hypotheses regarding protein function [1, 2]; however, most proteins are regulated by a complex array of post-translational events that may or may not be reflected in gene expression signatures [3, 4]. The resulting variability of the full compliment of proteins encoded by the genome (the proteome) renders the assignment of protein function via genomic means a virtually impossible task. Proteomic methods such as liquid chromatography–mass spectroscopy (LC-MS) platforms for shotgun analysis [5, 6], yeast two-hybrid methods [7, 8] and protein microarrays [9] have all contributed to the understanding of expression patterns, interactions, and in vitro functions of proteins. However, these methods are limited in that they do not address the activity of proteins in their native state. As it is the activity of a protein, rather than its expression level, that determines its role in cellular (patho)physiology, methods to directly interrogate the activity of natively expressed protein families amongst a background of high biological complexity have been developed. One such method, termed "activity-based protein profiling" (ABPP), has been systematically explored and reviewed, and has evolved into a standard tool for the identification and functional characterization of individual proteins in native proteomes consisting of tens of thousands of proteins [10–14].The cornerstone and hallmark of ABPP is the design of active-site directed small molecule probes capable of covalent protein modification. This design is crucial to the success of the ABPP experiment. These probes typically target a manageable fraction of the proteome with shared catalytic features. Once covalent modification of the protein has been achieved, a reporter group on the probe allows for visualization and/or separation of the labeled proteins via a number of methods. Thus, the application of ABPP results in fractionation of a proteome on the basis of catalytic activity. In this introductory chapter, we will highlight the basic principles of ABPP, as well as introduce the reader to the subsequent chapters of this volume.

2 Principles of ABPP

In this section, we introduce the basic principles of ABPP, including probe and tag design, briefly touch upon analytical platforms and classes of enzymes targeted for ABPP. We will then highlight some recent developments in the use of ABPP. The basic ABPP experiment is shown in Fig. 1.

In this experiment, the probe (consisting of a binding group, a linker, and a reporter group) is incubated with crude cell lysates or, if the probe is cell-permeable, with whole cells. The binding group specifically interacts with enzyme subsets and forms a covalent linkage. The now-labeled proteome can be analyzed by a variety of means, as discussed in Sect. 3.4. As these proteins have been

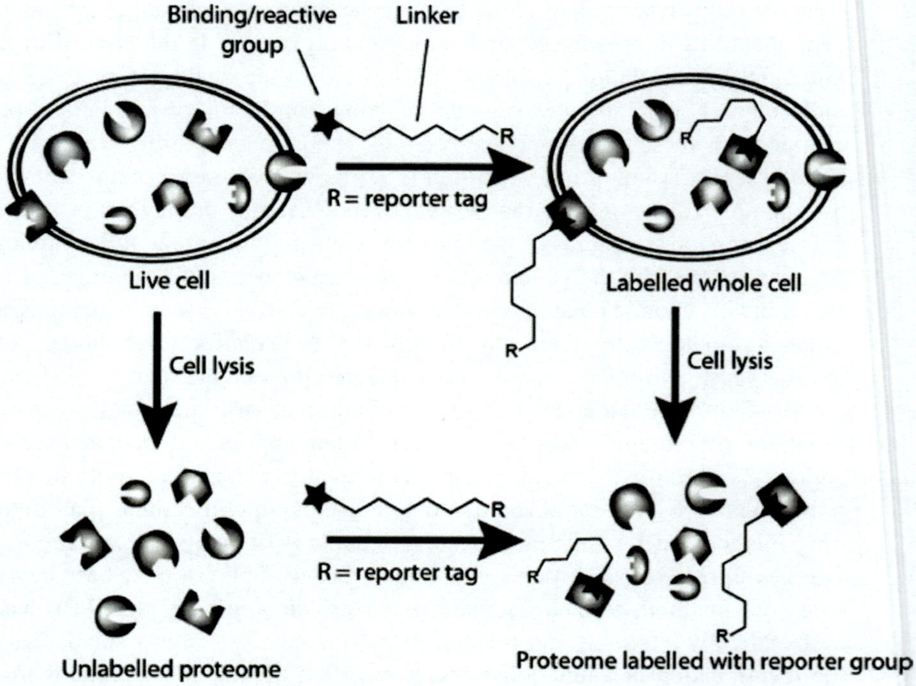

Fig. 1 Basic ABPP experiment

identified via the binding of a small molecule, the protein targets addressed by ABPP can be regarded as drugable, which is an important prerequisite for further medicinal consideration.

2.1 Probe and Tag Design

The design of specific probes is a crucial part of all ABPP experiments. ABPP probes typically possess three structural elements: (a) a binding group that, by virtue of noncovalent protein interactions, will guide the probe to specific enzymatic active sites; (b) a reactive moiety capable of covalently modifying enzymatic active sites in the proteome; and (c) an analytical handle used for visualization or separation of the probe-labeled protein. The reactive moiety of an ideal ABPP probe would exhibit reactivity towards proteins that would result in a small subset of proteins (tens to hundreds) being labeled. Should the probe be too reactive, many enzymes would be labeled at residues that are not involved in the enzyme function, resulting in corruption of the output data from the experiment [15–17]. Conversely, if the probe is not reactive enough, essential enzymes might go unlabeled. Reactive groups are generally based upon several chemical scaffolds, including mechanism-based inhibitors [18], protein-reactive natural products

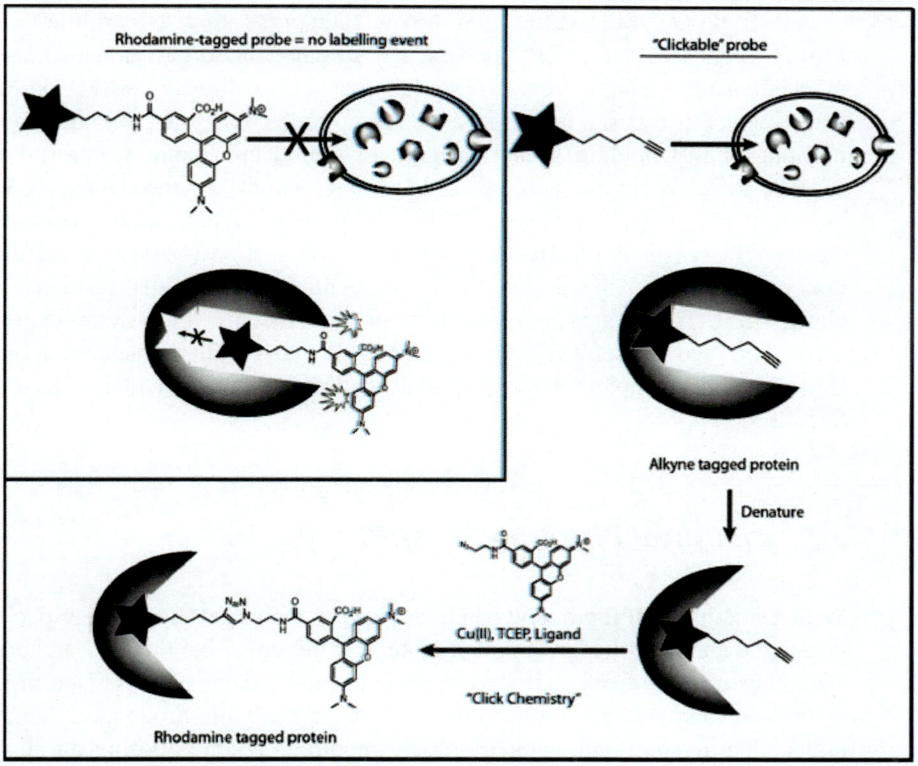

Fig. 2 Advantages of "click chemistry" as applied to ABPP

[13, 19], general electrophilic moieties [20–22], and photoreactive chemotypes [23, 24]. The binding group must, as in medicinal chemistry, present an optimal blend of noncovalent interactions, size, and hydrophobicity to allow for effective protein interaction in a native environment [25]. Finally, the analytical handle must be chosen for visualization, identification, or separation as needed, while not overly perturbing the noncovalent protein–probe interactions of the binding group or affecting cell permeability. This concern has largely been circumvented by the use of a sterically benign latent tag, such as an alkyne or azide that can be modified with a number of reporter groups via bio-orthogonal reactions such as the Cu(I)-catalyzed Huisgen [3+2] cycloaddition (click chemistry) [26–28] or the Staudinger ligation [29, 30]. These reactions serve to decouple the reporter or separation tag from the probe, allowing for less binding interference from the analytical handle and more flexibility in the choice of reporter tags available for attachment (Fig. 2).

As click chemistry requires catalytic copper, which is toxic to cells, in vivo applications have so far been restricted. To overcome this limitation, Bertozzi and coworkers introduced a second generation of click chemistry that is based on a strain-promoted reaction between a cyclooctyne and an azide moiety [31]. This reaction has been further improved over the past years and is a hot topic in ligation chemistry [32, 33].

ABPP probes fall within two broad categories: directed and nondirected probes [10]. Directed ABPP probes are designed to target enzymes within a mechanistically related family. Two approaches to directed ABPP probes are using mechanism-based inhibitors as reactive groups or alternatively incorporation of high-affinity binding groups specific to shared enzymatic features. For this strategy to be successful, a degree of knowledge of enzyme mechanism, structure, or ligand preferences is necessary [34]. Nondirected ABPP probes contain mild electrophiles or photoreactive groups and a variety of structural features that can noncovalently bind the probes to a variety of mechanistically distinct enzyme classes [35]. This strategy requires little or no knowledge of enzyme structure or active site, and is useful for extending ABPP to less well-characterized enzymes [36]. Several examples of directed and nondirected probes will be discussed in Sect. 2.3.

2.2 *Analytical Platforms for ABPP*

With a suitable ABPP probe in hand, labeled protein targets need to be specifically visualized and identified in complex proteomic mixtures. For this, researchers have a number of analytical platforms at their disposal. Each of these platforms has distinct advantages and disadvantages in the criteria of throughput, sensitivity, target identification, and sample amount required. When choosing an analytical method, these criteria must be considered in conjunction with the specific experimental problem in order to maximize the information obtained. In this section, we briefly discuss the major analytical platforms available to the researcher and highlight the advantages and disadvantages of their use.

2.2.1 Gel-Based Platforms

Gel-based analytical methods are the most mature standard technique for ABPP [37–39]. They have the advantages of being simple, robust, fast, and amenable to high-throughput analysis of potentially hundreds of proteomes per day. In this technique, probe-labeled proteomes (~30–50 μg) are applied to an SDS polyacrylamide gel and resolved by one- (1D) or two-dimensional (2D) electrophoresis. The readout is visualized by either in-gel fluorescence scanning or avidin blotting, for fluorescent and biotinylated ABPP probes, respectively. However, this readout does not reveal the identity of the labeled target enzymes. Protein identity is typically revealed by cutting out the visualized bands, followed by in-gel trypsin digest and tandem mass spectroscopy (MS/MS) analysis. Furthermore, a biotin–(strept)avidin enrichment step is often required before applying the labeled proteomes to the gel in order to reduce background levels of nontarget proteins and provide enough target protein for accurate MS analysis. Other disadvantages of gel-based analytical platforms include low resolution and a low sensitivity of about 10 pmol/mg proteome, which is sometimes below the detection limit of physiologically relevant enzymes [40, 41]. In the commonly used 1D gel electrophoresis, proteins are separated on the basis of their molecular weight. Homologous enzymes tend to have a similar molecular mass, so members of a protein family that reacts with an ABPP probe will tend to co-migrate on the gel, obscuring or confusing the signal and subsequent MS analysis. Despite these limitations, the advantages of gel-based platforms make this technique extremely useful and still commonly employed [42, 43].

2.2.2 LC-MS-Based Platforms

With the principal limitations of gel-based platforms being resolution and the need for a secondary identification step, multiple LC-MS strategies for analyzing probe-labeled proteomes have been developed. The conceptually simplest of these methods is multidimensional protein identification technology (MudPIT)-ABPP [6, 44]. In this technique, proteomes are labeled with a biotinylated ABPP probe. The labeled proteome is then incubated directly with (strept)avidin beads, and the supernatant filtered off. Enriched proteins are then subjected to on-bead tryptic digest and the resulting peptides are analyzed by multidimensional LC-MS/MS. A database search of these peptide hits using the SEQUEST algorithm then reveals the identity of the protein target (Fig. 3).

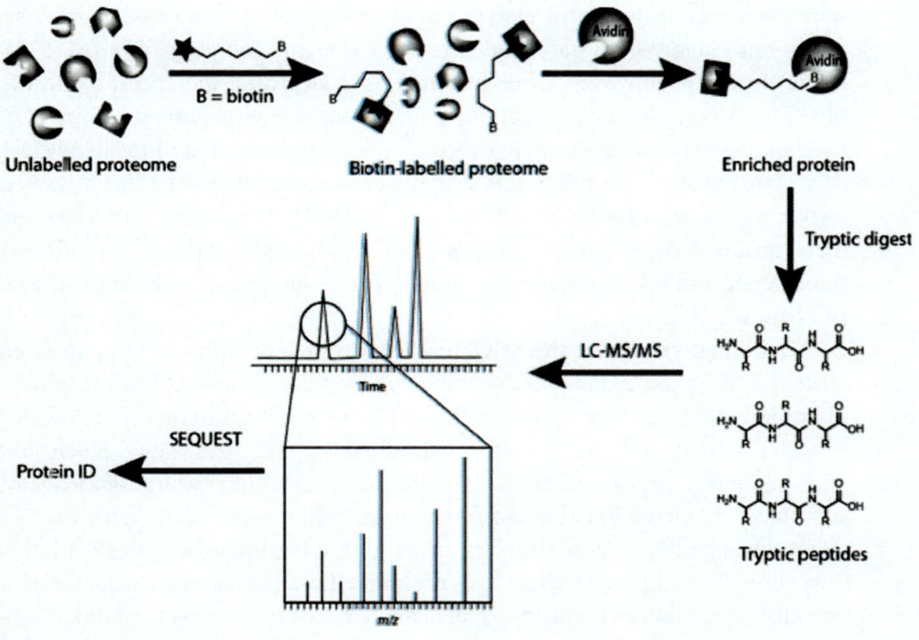

Fig. 3 Basic MudPIT ABPP

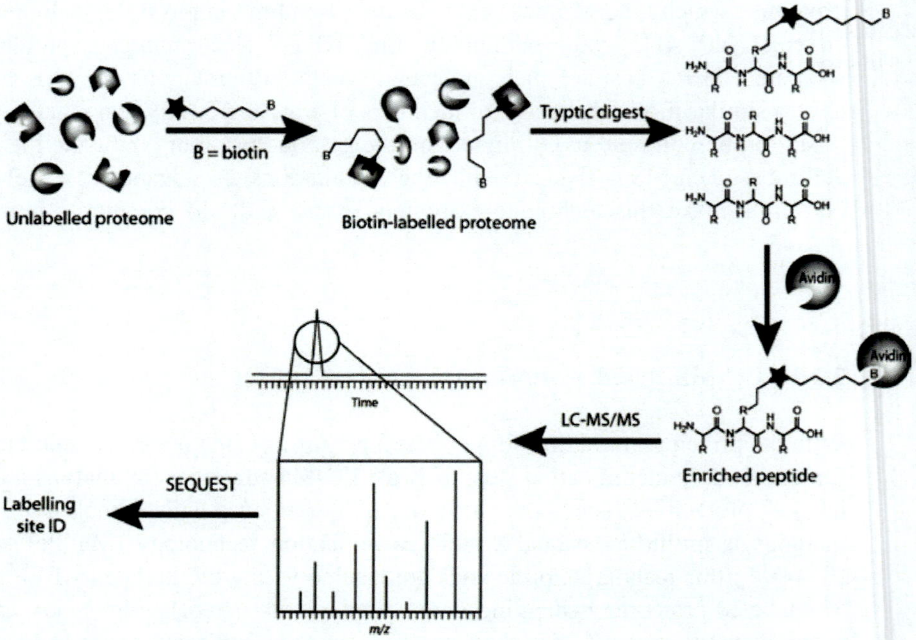

Fig. 4 Active site peptide profiling (ASPP)

Benefits of this technique are exceptional resolution and sensitivity (1 pmol/mg proteome), as well as target identification. Additionally, MudPIT-ABPP can determine the relative levels of enzyme activities in two or more proteomic samples using semiquantitative parameters such as spectral counting [45, 46]. This technique does not, however, offer information concerning the specific site of probe labeling. A complementary technique called active site peptide profiling (ASPP) is used to reveal this information. This method is similar to that of the MudPIT-ABPP described above, except that in ASPP, the probe-labeled proteomes are subjected to tryptic digest prior to the enrichment step [47, 48]. The probe-labeled peptides are then enriched using either (strept)avidin or antibody resins for biotinylated or fluorescent probes, respectively, eluted from the resin, and then analyzed by LC-MS/MS (Fig. 4).

A modified version of the SEQUEST algorithm is employed to identify enzyme targets. Furthermore, database searches for conserved nucleophilic residues in the identified peptides can confirm these peptides as catalytically relevant. Both MudPIT-ABPP and ASPP offer complimentary sets of data. MudPIT-ABPP analyzes many peptides from the tryptic digest, thus increasing the probability of accurately identifying and quantifying target proteins. In contrast, the use of ASPP strengthens confidence that probes are targeting legitimate active sites rather than unspecific labeling. In an effort to consolidate these two sets of data and make them available from one experiment, Speers and Cravatt developed the tandem orthogonal proteolysis (TOP)-ABPP method [49]. In this platform, a tobacco etch virus

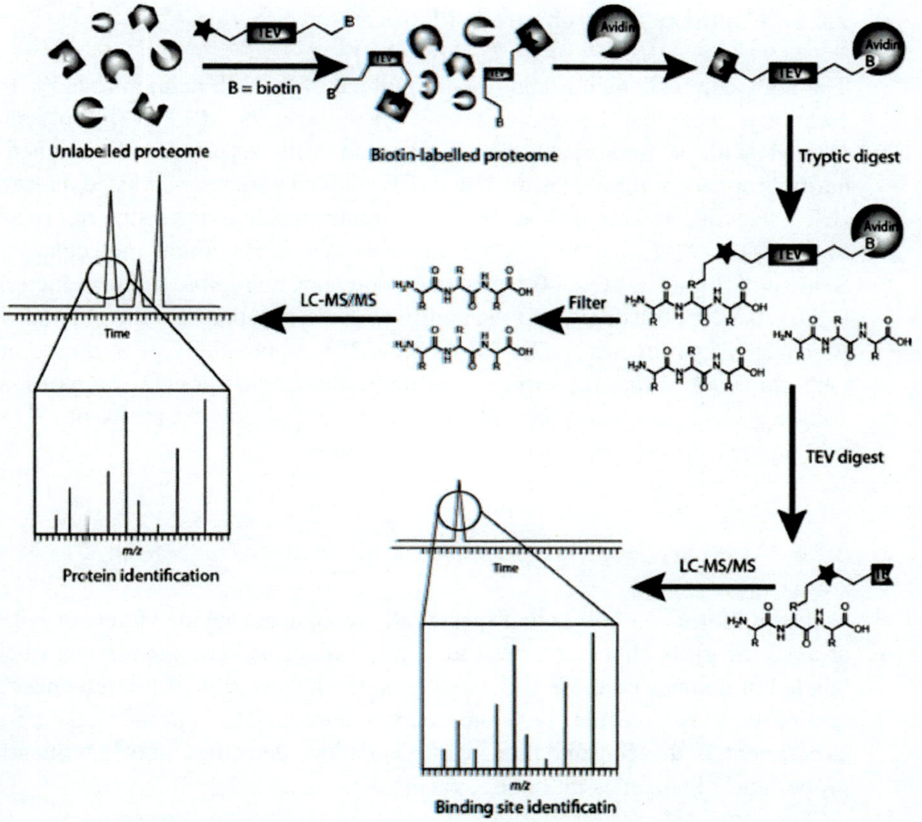

Fig. 5 Tandem orthogonal proteolysis (TOP) ABPP. *TEV* Tobacco etch virus

(TEV) protease cleavage site is inserted between the binding/reactive group and the analytical handle. Labeling of proteomes with these probes followed by (strept) avidin resin incubation yields an enriched proteome. Trypsin digest followed by filtration yields a supernatant that can be analyzed by multidimensional LC-MS/MS analysis as in the MudPIT-ABPP experiment above. In the last step, the probe-labeled active site peptides are released from the resin by cleavage with TEV protease and subjected to MudPIT analysis (Fig. 5).

The two separate LC-MS/MS runs of the tryptic digest and TEV protease digest can then be analyzed in tandem, offering a second dimension of validation for the protein targets and their catalytic binding sites. For example, protein hits found in the TEV elution, but not in the tryptic digest can be regarded as false assignment. Protein hits found in the tryptic digest, but not in the TEV elution can be regarded as background (nonspecific) hits. While the LC-MS methods outlined above are extremely powerful, they are difficult to perform in parallel, require large amounts of proteome (~1 mg) and are time consuming (several hours per sample) and thus suffer from low throughput.

2.2.3 Capillary Electrophoresis Platforms

The high sample demands and low-throughput of LC-MS methods have led to the creation of a capillary electrophoresis (CE) platform for ABPP [48]. Proteomes are labeled with a fluorescent probe, digested with trypsin, and enriched with antifluorophore antibody resins. Use of CE coupled with laser-induced fluorescence (LIF) detection to analyze the enriched peptides resulted in far superior resolution to 1D SDS-PAGE, particularly for enzymes that share similar molecular masses. Sensitivity limits of ~0.05–0.1 pmol/mg proteome, negligible sample requirements (~0.01–0.1 μg proteome), and the ability to perform rapid CE runs in parallel with 96-channel instruments, make CE-based ABPP a potentially powerful technique. One drawback is that the identities of the probe-labeled proteins are not immediately apparent, and correlated LC-MS experiments must be performed to assign protein identities to the peaks on the CE readout.

2.2.4 Microarrays

In this platform, antibodies that specifically recognize enzyme targets of ABPP are arrayed on glass slides and used as capture reagents. The application of probe-labeled proteomes on these slides results in the localization of fluorescence, which can be directly detected by fluorescence scanning [50]. The end result for this experiment is the consolidation of the isolation, detection, and identification of probe-labeled enzymes into one step (Fig. 6).

The sensitivity of this platform is excellent (1 pmol/mg proteome) and sample consumption is minimal (~1–3 μg proteome). Moreover, because the detection and identification of probe-labeled targets is performed in one step, ABPP microarrays are amenable to high-throughput proteomic experiments. The key drawback to this method is the relative lack of high-quality commercially available antibodies that can selectively recognize enzyme targets [51]. An earlier precursor to this experiment was performed by Harris and coworkers whereby a series of actylate-based ABPP probes were covalently attached to a peptide nucleic acid (PNA) tag. Incubation of these PNA-encoded probes with a proteome results in some probes reacting with cysteine proteases [52]. Gel filtration to remove unreacted probes,

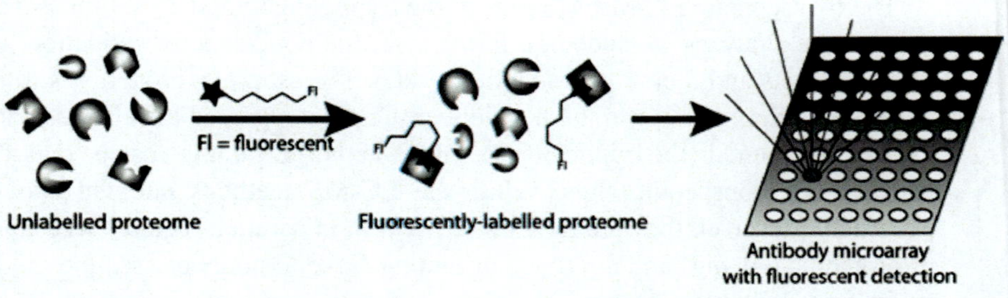

Fig. 6 Microarray-based ABPP

followed by hybridization to an oligonucleotide microarray gives a readout indicating which probes are interacting with the proteome.

2.3 Enzyme Classes Addressable by ABPP

The design of directed activity-based probes requires the selection of a suitable reactive group to interact with the enzyme class of interest. Ideally, the probe should form covalent linkages to a desired protein family while remaining inert to all other species within the proteome (and cell). This often requires knowledge of the design of mechanism-based inhibitors and affinity labels for these enzyme classes. Many of the best directed activity-based probes are simply well-characterized covalent inhibitors attached to an analytical handle. In this section, we will provide a brief survey of some of the enzyme classes that can be interrogated using directed ABPP approaches. This survey is not comprehensive; rather we have tried to provide a broad overview of directed ABPP.

2.3.1 Serine Hydrolases

The serine hydrolase family is one of the largest and most diverse classes of enzymes. They include proteases, peptidases, lipases, esterases, and amidases and play important roles in numerous physiological and pathological process including inflammation [53], angiogenesis [54], cancer [55], and diabetes [56]. This enzyme family catalyzes the hydrolysis of ester, thioester, and amide bonds in a variety of protein and nonprotein substrates. This hydrolysis chemistry is accomplished by the activation of a conserved serine residue, which then attacks the substrate carbonyl. The resulting covalent adduct is then cleaved by a water molecule, restoring the serine to its active state [57] (Scheme 1).

The greatly increased nucleophilicity of the catalytic serine distinguishes it from all other serine residues and makes it an ideal candidate for modification via activity-based probes [58]. Of the electrophilic probe types to profile serine hydrolases, the fluorophosphonate (FP)-based probes are the most extensively used and were first introduced by Cravatt and coworkers [38, 39]. FPs have been well-known inhibitors of serine hydrolases for over 80 years and were first applied as chemical weapons as potent acetylcholine esterase inhibitors. As FPs do not resemble a peptide or ester substrate, they are nonselective towards a particular serine hydrolase, thus allowing the entire family to be profiled. FPs also show minimal cross-reactivity with other classes of hydrolases such as cysteine-, metallo-, and aspartylhydrolases [59]. Furthermore, FP-based probes react only with the active serine hydrolase, and not the inactive zymogen, allowing these probes to interact only with functional species within the proteome [59]. Extensive use of this probe family has demonstrated their remarkable selectivity for serine hydrolases and resulted in the identification of over 100 distinct serine hydrolases

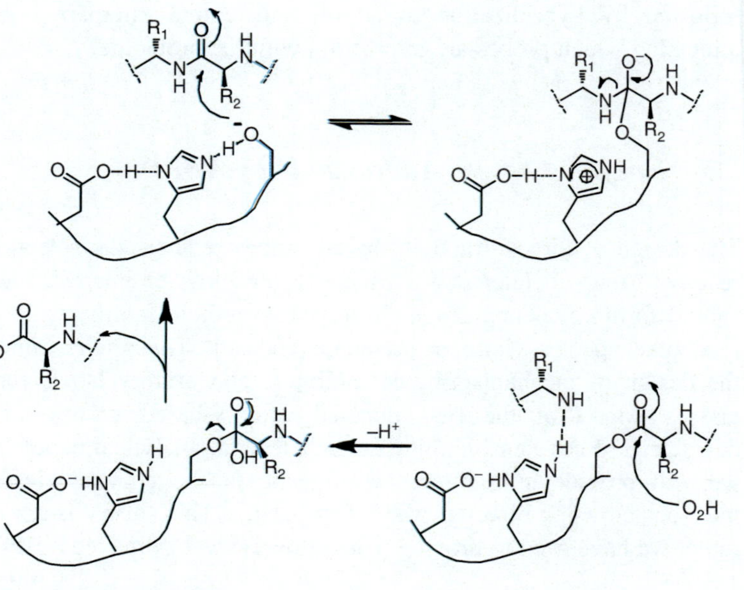

Scheme 1 Mechanism of action of serine hydrolases

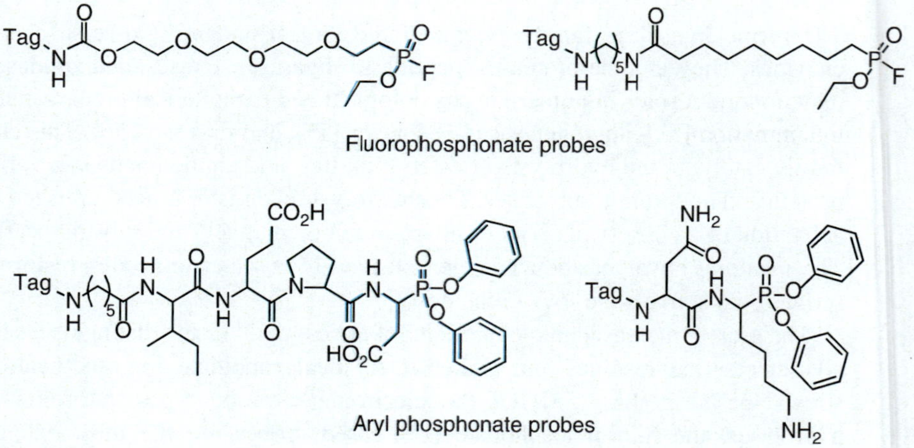

Fig. 7 Structures of fluorophosphonate and arylphosphonate probes

including a large number of previously uncharacterized enzymes with respect to their endogenous substrates and physiological function [38, 39, 44, 48, 60, 61]. Complementing the broad-spectrum activity of FP probes, a series of arylphosphonates developed by Craik and Mahrus display less promiscuity towards serine hydrolases, and substrate-mimetic chemical scaffolds can promote specific interactions with granzyme A and B [62] (Fig. 7). This example demonstrates that the reactivity of promiscuous, highly active probes can be finely tuned towards individual members of enzyme superfamilies.

2.3.2 Cysteine Proteases

Cysteine proteases are a large class of enzymes that are involved in multiple physiological and pathological events such as bone remodeling [63], cancer invasion [64, 65], malaria [66], and arthritis [67]. Similar to serine hydrolases, cysteine proteases possess a conserved cysteine residue of heightened nucleophilicty via an adjacent histidine residue. Attack of the thiolate on the substrate carbonyl results in an intermediate thioester that is readily cleaved by water to generate the peptide product and regenerated cysteine protease [57]. Many diverse electrophilic activity-based probes have been developed for labeling the cysteine protease family, including epoxides [68, 69], vinyl sulfones [70], diazomethyl ketones [71], α-halo ketones [72], and acyloxymethyl ketones (AOMKs) [73]. Of these examples, the AOMK probes have demonstrated exceptional class-wide reactivity and selectivity with cysteine proteases [74]. AOMK probes derive their high selectivity for cysteine proteases due to their relatively complex chemical behavior (Scheme 2).

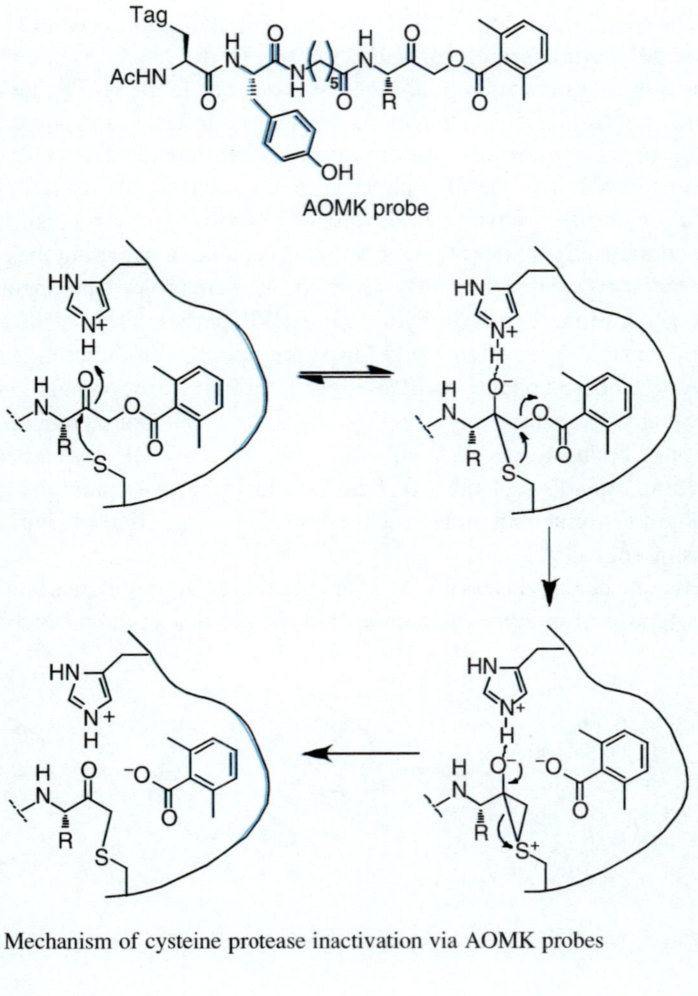

Scheme 2 Mechanism of cysteine protease inactivation via AOMK probes

Following attack of the cysteine thiolate to the probe ketone group, the resulting sulfide accelerates the displacement of the carboxylate leaving group, resulting in the formation of an episulfonium intermediate. Finally, opening of the sulfonium intermediate generates a stable thiomethyl ketone adduct [75]. Other cysteine protease ABPP probes, by virtue of their electrophilicity, exhibit labeling of many hydrolase enzyme classes. Using of a library of AOMK probes, researchers have targeted numerous members of the two major classes of cysteine proteases, the CD (caspase-3, legumain, arginine- and lysine-gingipains) and the CA (cathepsin B and cathepsin L) clans [76, 77].

2.3.3 Metallohydrolases

Metallohydrolases are the third major class of hydrolytic enzymes. Unlike serine or cysteine hydrolases, the metallohydrolases achieve substrate hydrolysis via a zinc-activated water molecule [78–80] (Scheme 3).

The metalloproteases (MPs) and matrix metalloproteinases (MMPs) are a class of metallohydrolases of particular interest to the pharmaceutical industry due to their role in a number of pathological processes [81–83]. The lack of an enzyme-bound nucleophilic residue in the metallohydrolases complicates the design of ABPP probes for this class of enzymes. Rather than mechanism-based and electrophilic probes for ABPP, photoreactive variants of reversible inhibitors of metallohydrolases have been developed [84–86]. These reversible inhibitors usually contain a hydroxamate moiety that is capable of chelating the catalytic zinc ion in a bidentate manner [79, 80]. The hydroxamate moiety was incorporated into the first generation of metallohydrolase ABPP probes along with a benzophenone group capable of covalent bond formation upon UV irradiation (Scheme 4).

A probe based on the MMP inhibitor Ilomastat was reported by Saghatelian et al. as capable of selectively labeling the active, but not zymogen or endogenous inhibitor-bound forms of MMPs [85] (Fig. 8). This selective labeling allowed the functional analysis of the MP family, a task that was previously inaccessible to standard genomic and proteomic methods due to the rigid in vivo regulation of this class of enzymes.

Histone deacetylases (HDACs) are the second major class of metallohydrolases. This family of enzymes is responsible for the deacetylation of lysine residues in

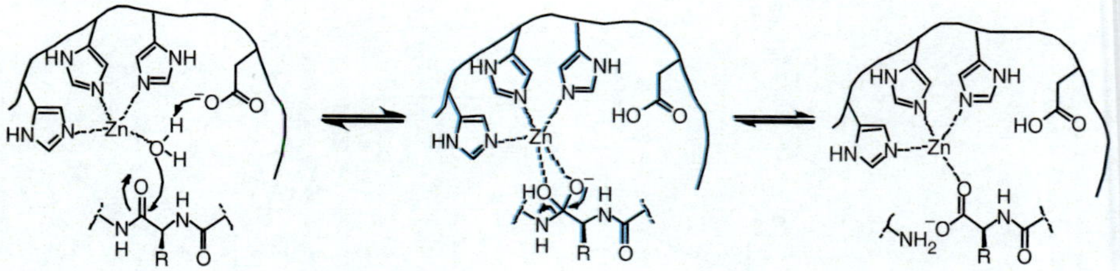

Scheme 3 Metallohydrolase mechanism of action

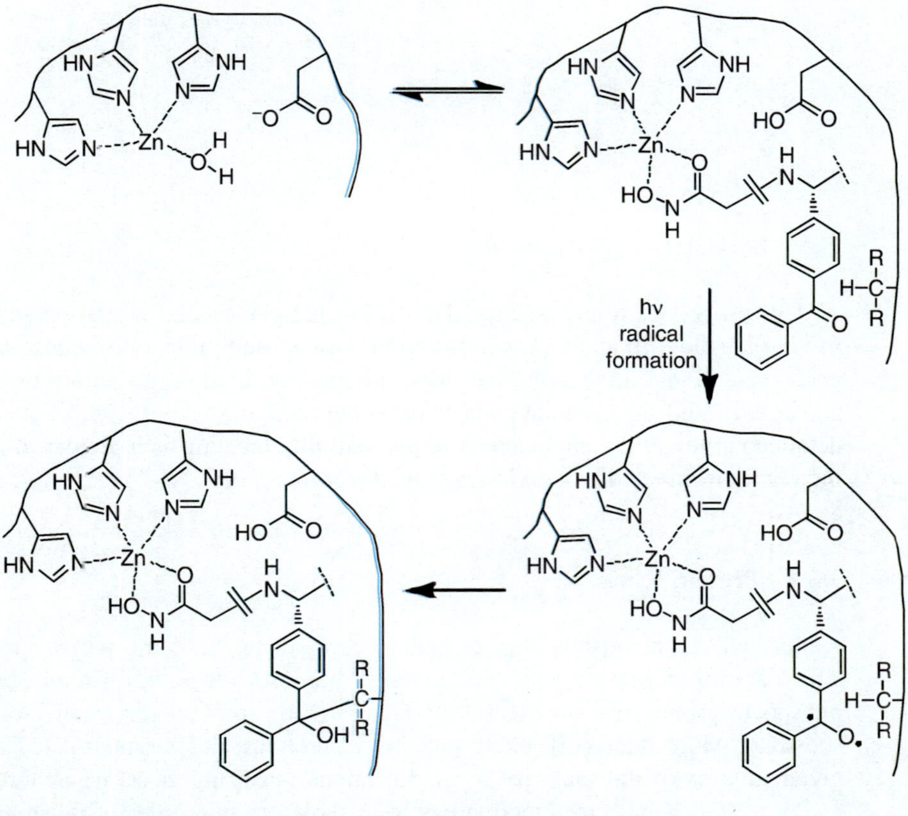

Scheme 4 Metallohydrolase labeling with hydroxamate photoprobes

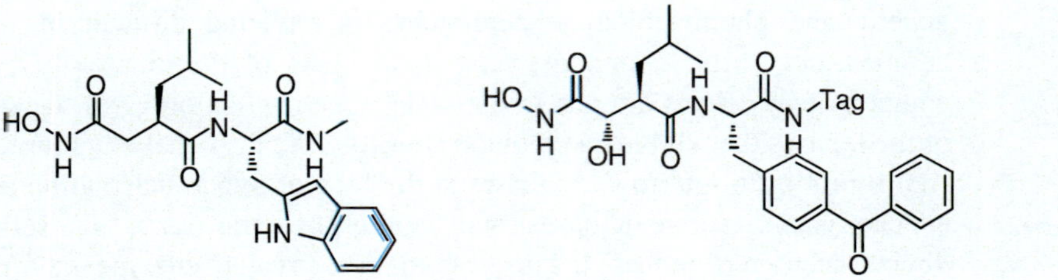

Ilomastat

Fig. 8 Ilomastat and its hydroxamate ABPP probe mimic

histone proteins. The reversible acetylation and deacetylation of these residues plays a central part of transcriptional activation and repression [87], and inhibitors of some HDACs have been reported to possess potent anticancer activity [88, 89]. One such inhibitor is suberoylanilide hydroxamic acid (SAHA), which has been approved by the FDA for the treatment of cutaneous T cell lymphoma [89]. Based on SAHA, Salisbury and Cravatt synthesized a mimic ABPP probe, SAHA-BPyne (Fig. 9).

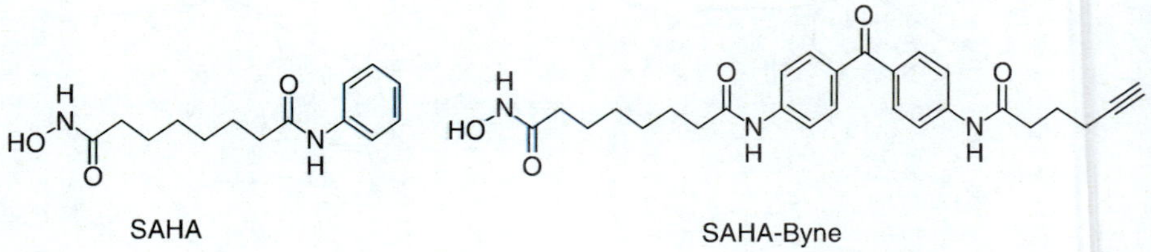

Fig. 9 SAHA and its ABPP probe mimic

This probe was found to target HDACs in proteomes, and was also capable of in vivo labeling. In addition, several HDAC-associated proteins were also labeled by SAHA-BPyne, indicating that these proteins are in close proximity to HDAC active sites and may regulate substrate recognition and activity [90]. For a more detailed survey of the applications of photoaffinity labeling as it applies to ABPP, the reader is directed to a review by Overkleeft et al. [91]

2.3.4 Protein Kinases

Almost all signal transduction cascades are mediated in some part by kinases, which are the largest enzyme class in eukaryotic proteomes [92]. Kinases transfer phosphate groups from nucleotide cofactors to the protein (or small molecule) substrate. More than 30% of all eukaryotic proteins are phosphorylated at any given time, with the majority of modifications occurring at serine or threonine residues [93]. Kinase dysfunction has been shown to play roles in inflammation, cancer, and diabetes [94] and kinases are promising targets for therapeutic intervention [95, 96]. However, functional assignment of kinases by conventional genetic and pharmacological approaches is rendered difficult by system-wide compensatory effects amongst related members of the kinases [97, 98]. These compensatory effects are due to the high sequence homology of the active region of the kinases that all bind a common substrate, ATP. A challenge for the design of ABPP probes to interrogate kinases is the lack of hypernucleophilic residues that are catalytically active. Kinases catalyze phosphate transfer to their substrates via a direct transfer mechanism that does not involve covalent enzyme intermediates and, as such, their active sites typically do not contain any unusually nucleophilic residues. However, sequence comparisons have shown that almost all protein kinases have at least one conserved lysine residue within their active sites, representing a possible modification site for an appropriate probe [92, 99]. These lysines residues are positioned close to β- and γ-phosphates of bound ATP, providing electrostatic interactions with the phosphate backbone [100]. It was then reasoned that kinase ABPP probes based on ATP might take advantage of the conserved lysine. In an approach by Patricell et al., a biotinylated acyl group was placed at the terminal phosphate group of ATP or ADP [101]. These acyl phosphates display reactivity with amines while displaying stability in aqueous solutions (Scheme 5).

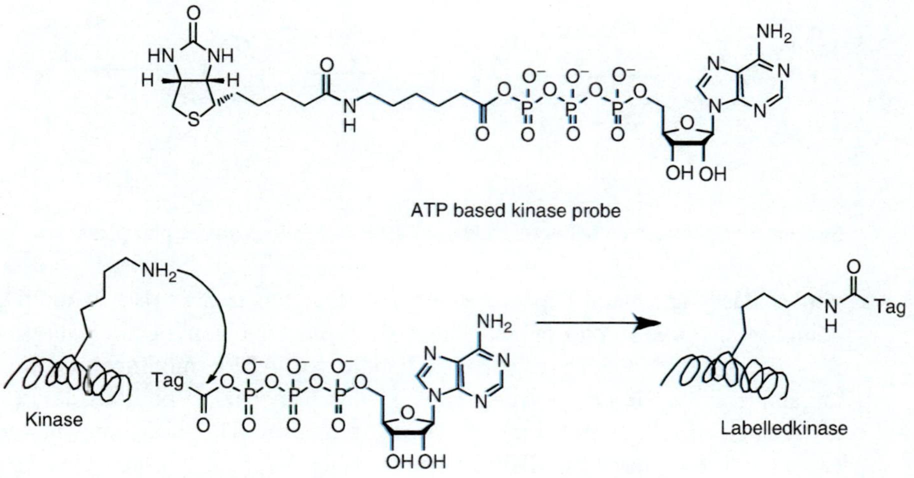

Scheme 5 ATP-based kinase probes

The probes of this type were shown to selectively label at least 75% of human kinases in crude cell lysates, thus demonstrating their selectivity and promiscuity for kinases [101]. As a follow up, the labeled kinases were subjected to proteolytic digestion, and the biotinylated peptides purified on avidin beads and analyzed by LC-MS/MS. This analysis demonstrated that the site of probe labeling was indeed the conserved active-site lysine as predicted. In contrast to the promiscuity demonstrated by the acyl phosphate probes, several selective covalent inhibitors of protein kinases have been used as ABPP probes. Wortmannin is a natural product derived from the fungus *Penicillium funiculosum*. It is a potent and specific covalent inhibitor of phosphoinositide 3-kinase (PI3K) and the PI3K-related kinase (PIKK) families [102, 103]. The use of natural products in relation to ABPP is covered by Breinbauer et al. [104].

2.3.5 Phosphatases

While protein kinases are responsible for the phosphorylation of their substrates, protein phosphatases perform the opposite duty, removing phosphate groups from their substrates, thus countering the functional impact of the kinases. The two major types of protein phosphatases are the serine/threonine phosphatases and tyrosine phosphatases. Several natural compounds with potent serine/threonine phosphatase inhibitory activity have been identified, including the cyanobacterial metabolite microcystin [105, 106]. This compound labels its targets via a Michael addition of a noncatalytic active site cysteine residue with an acceptor in the macrocyclic peptide backbone [107]. A fluorescent probe based on microcystin was synthesized by Shreder et al., and its use in Jurkat lysates identified two previously undescribed phosphatase targets of microcystin, PP-4 and PP-5 [108]. Whereas serine/threonine

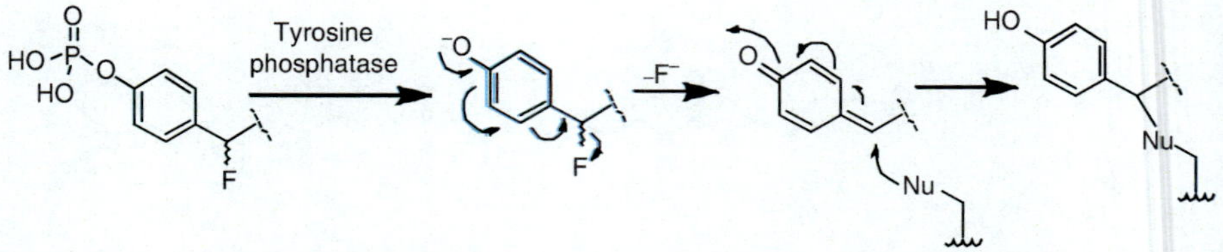

Scheme 6 Activation of 4-fluoromethylaryl ABPP probes for tyrosine phosphatases

phosphatases are metal-dependent enzymes that function as part of multisubunit complexes, tyrosine phosphatases use a cysteine nucleophile for catalysis. The presence of this residue suggests a handle for ABPP modification; however, the active site cysteine of tyrosine phosphatases appears to be much less reactive towards electrophiles than that of cysteine proteases. This has somewhat complicated the development of ABPP probes for tyrosine phosphatases. However, some progress towards this goal has been made. Lo et al. made initial probes for tyrosine phosphatases, which were based on a 4-fluoromethylaryl phosphate moiety [109]. Tyrosine phosphatase-mediated hydrolysis of the phosphate moiety generates a *p*-quinone methide intermediate that can react with an active site nucleophile (Scheme 6).

Although these probes show potent, heat-sensitive labeling profiles with purified tyrosine phosphatases, suggestive of specific active site modification, their use in complex proteomes was not reported [109]. Zhang and coworkers introduced a more specific class of tyrosine phosphatase probe [110]. This probe consists of an α-bromobenzylphosphonate moiety that acts as a tyrosine phosphate mimic and a mechanism-based inhibitor of tyrosine phosphatases. The chemistry of the probe mode of action is shown in Scheme 7.

Although these probes reacted specifically with tyrosine phosphatases, low water stability and limited cell permeability hindered their use in further ABPP studies [110]. The same group subsequently introduced a new class of phosphatase probes based on an aryl vinyl sulfone moiety [35].

2.3.6 Glycosidases

Glycosyltransferases and their counterparts, glycosidases, are the major enzymes for the formation and hydrolysis, respectively, of glycosidic bonds. The glycosidases can be divided into two classes according to enzymatic mechanism, i.e., retaining or inverting the anomeric center of the cleaved carbohydrate [111]. The β-retaining glycosidases are the mechanistic class of glycosidases that have received the most attention for activity-based proteomic studies (Scheme 8).

For many glycosidases, the active site is highly tailored, consisting of extensive interactions between substrate and enzyme. This tight binding complicates the design of ABPP probes as there is little remaining room for larger probes or tags. However, the β-glycosidases utilize a covalent intermediate, providing a

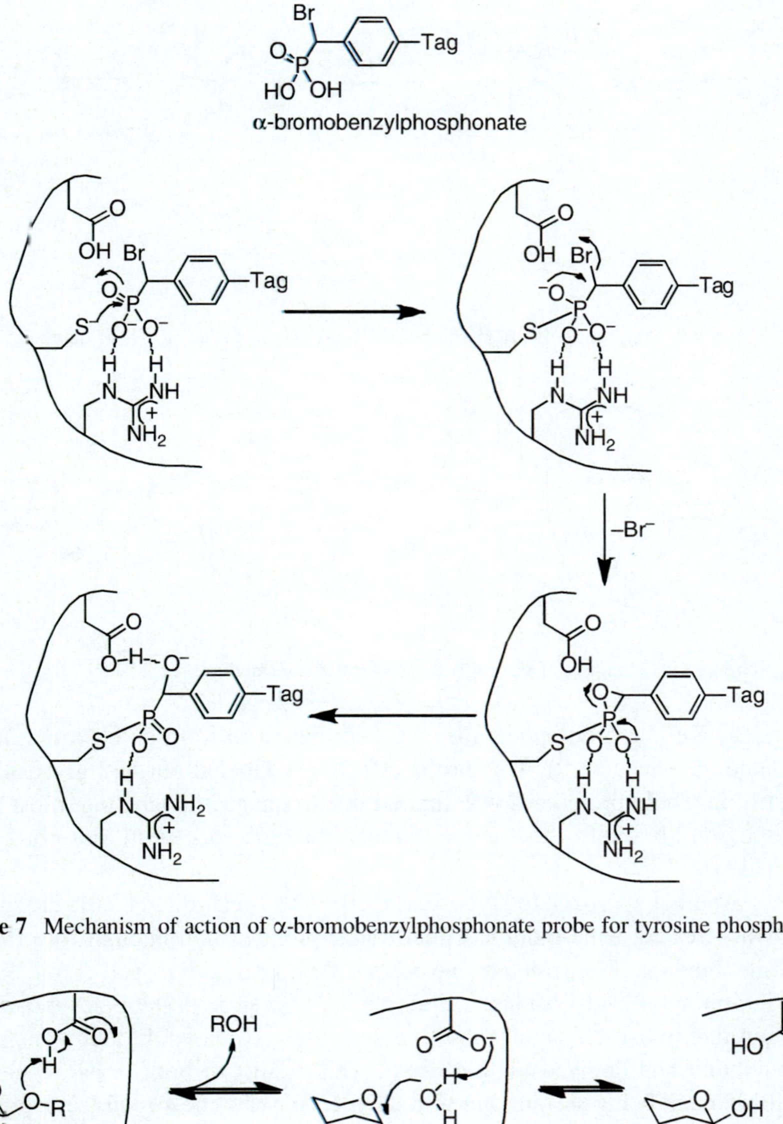

Scheme 7 Mechanism of action of α-bromobenzylphosphonate probe for tyrosine phosphatases

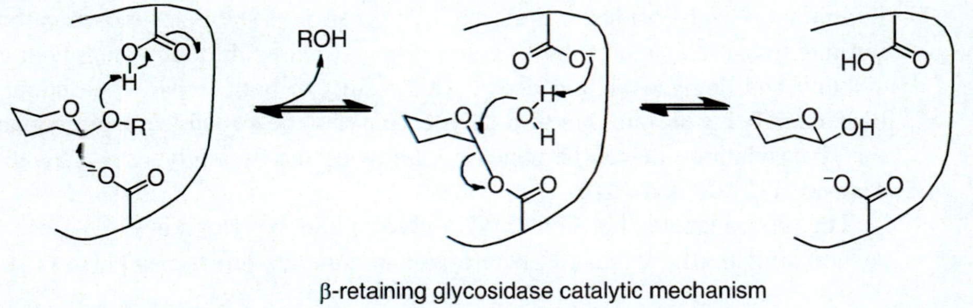

Scheme 8 Mechanism of action of β-retaining glycosidases

strategic handle for protein modification [111]. One of the first glycosidase ABPP probe designs was adapted from a previously introduced latent quinone methide reactive group used for tyrosine phosphatases (Sect. 2.3.5). Upon glycosidic bond

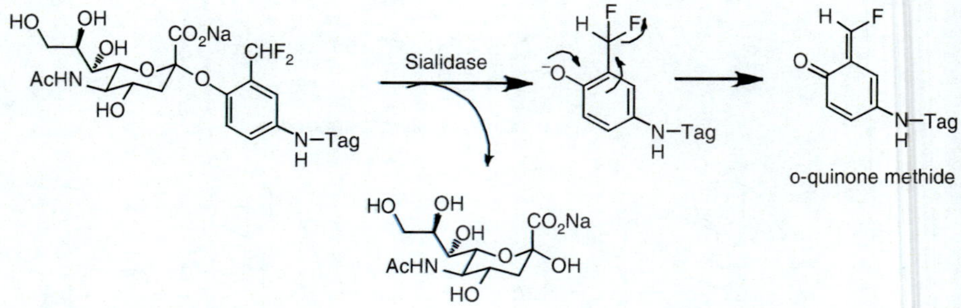

Scheme 9 Activation of glycosidase ABPP probe by cleavage of sialic acid cleavage

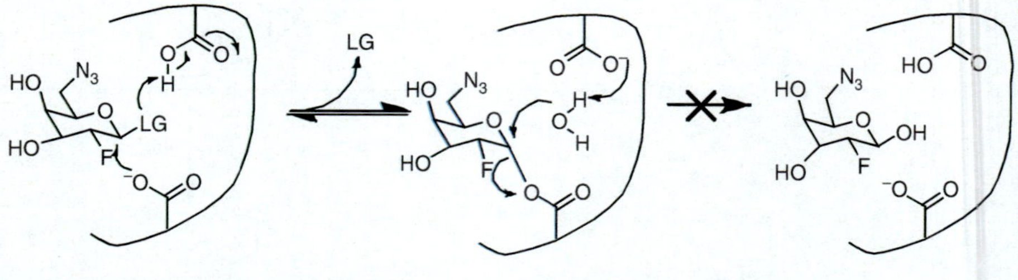

Scheme 10 Trapping of fluoroglycosyl-enzyme intermediate for ABPP

cleavage, the probe undergoes 1,6-elimination to form an electrophilic *o*-quinone methide (Scheme 9). This probe effectively labeled purified glycosidases, but its use in whole proteomes was limited due to unspecific labeling, most likely due to the equilibration of the active quinone methide species in the proteomic mixture [112].

Another approach to glycosidase ABPP was performed by the group of Vocaldo, who created azide-modified fluorosugar probes small enough for effective active site binding. Fluorosugars have been demonstrated to act as mechanism-based inhibitors of β-glycosidases. Fluorine incorporation destabilizes the oxacarbenium ion-like transition state of the β-retaining glycosidases through which both glycosylation and deglycosylation occur. This results in both steps of the double displacement being slowed, but through incorporation of a good leaving group at C-1, the glycosylation rate can be increased, allowing the fluoroglycosyl-enzyme to be trapped [113] (Scheme 10).

The probes labeled purified LacZ, a glycosidase expressed in *E. coli*, as well as several structurally unrelated glycosidases in complex proteomes [114, 115].

2.3.7 Cytochrome P450

The cytochrome P450 family of enzymes is a large and diverse family of monooxygenases whose function is to oxidize organic substrates. They perform roles in the metabolism of endogenous signaling enzymes, xenobiotics, and drugs [116].

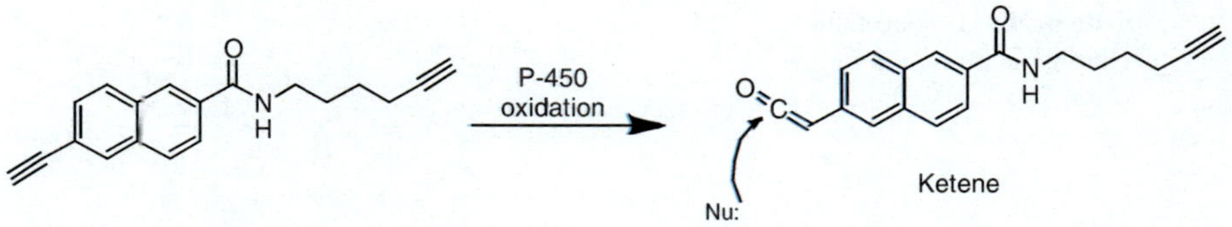

Scheme 11 Activation of an aryl alkyne to an electrophilic ketene by cytochrome P450

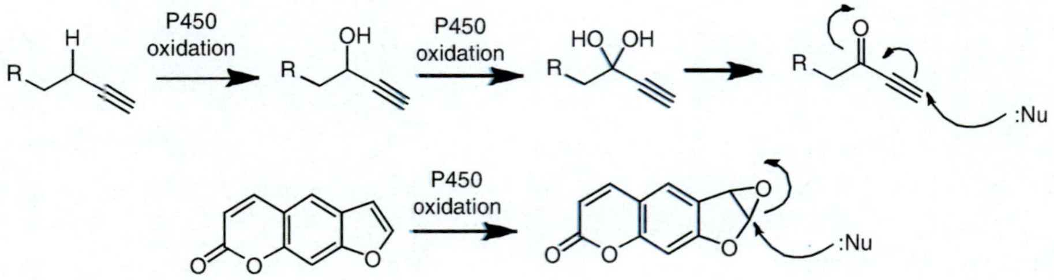

Scheme 12 Activation of propinyl and furanocoumarins to electrophilic moieties via cytochrome P450 oxidation

The human genome encodes 57 distinct P450 enzymes, but only five of these are responsible for 90% of the metabolism of current clinical drugs, leaving many members of this family uncharacterized [117]. The in vivo activity of P450 enzymes is tightly regulated by multiple factors: membrane composition and localization, protein-binding partners [118], post-translational modification [119], endogenous concentrations of cofactors, and regulatory enzymes. The well-studied mechanism-based inhibition of P450 enzymes by aryl acetylenes provided the basis for an ABPP probe introduced by Wright and Cravatt [120]. This probe has a 2-ethynylnapthalene core, where the naphthalene group directs the probe towards P450 active sites. There, oxidation of the aryl acetylene moiety results in a ketene intermediate that can then acylate nucleophilic residues within the enzyme active site (Scheme 11).

The alkyl acetylene that serves as a latent reporter group is not subject to the same enzymatic oxidation. This probe was shown to be capable of labeling multiple P450 enzymes both in vitro and in vivo [120]. This probe was also used to monitor induction and inhibition of P450 enzymes in living mice, providing a potentially powerful assay to characterize P450–drug interactions in vivo. In addition to the 2-ethynylnapthalene core, other ABPP probes that are capable of labeling P450 include the aryl alkyne, propinyl and furanocoumarin moieties [121], which upon P450 oxidation are capable of reaction with protein nucleophiles (Scheme 12).

2.3.8 Arginine Deiminases

Arginine deiminases catalyze the conversion of arginine residues to citrulline residues in proteins. Deregulation of protein arginine deiminase 4 (PAD4) is

Scheme 13 Haloacetamide arginine deiminase probes and modes of labeling

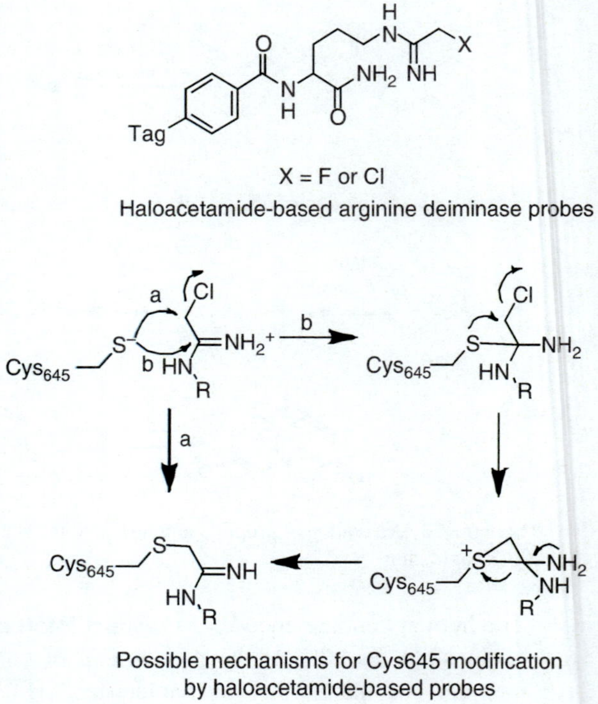

strongly implicated in increased intensity and occurrence of rheumatoid arthritis (RA) as well as other diseases including colitis and cancer [122, 123]. Because the activity of PAD4 is upregulated in RA, its inhibition represents a potential therapeutic target for the treatment of RA. To date, several haloacetamidine-based compounds have been studied as mechanism-based inhibitors of PAD4 [124, 125]. **These compounds inactivate PAD4 via the covalent modification of cysteine** 645 via one of two potential mechanisms, and only do so to the active, calcium-bound form of PAD4 [126] (Scheme 13).

By adapting these inhibitors to act as ABPP probes, Thomoson and coworkers have synthesized a series of probes that were capable of highly selective labeling of both PAD4 as well as PAD4 binding proteins in complex proteomes [127, 128].

2.3.9 The Proteasome

The proteasome is the primary protein degradation machinery in cells, and plays a central role in the degradation of abnormal and damaged proteins, cell cycle regulators, oncogens, and tumor suppressors [129]. The eukaryotic proteasome performs this process by catalyzing the hydrolysis of peptides using an activated nucleophilic threonine residue [130, 131]. As it relates to human disease, certain cancer cells are especially dependent on high levels of proteasome activity for survival, which has lead to the development of proteasome inhibitors for the

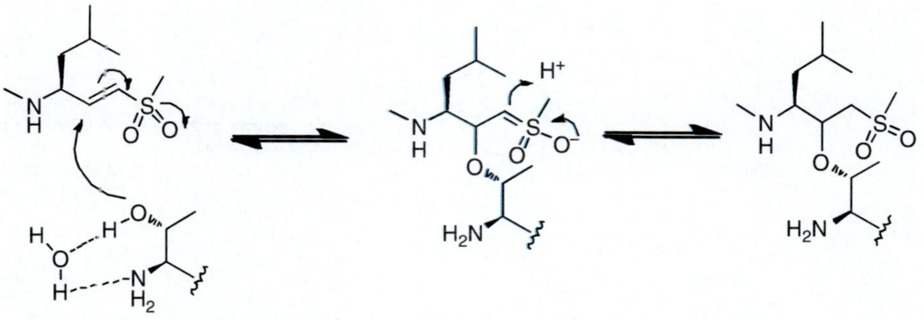

Scheme 14 Modification of proteasome threonine with vinyl sulfone ABPP probe

treatment of multiple myeloma [132]. To date, several classes of inhibitors of the proteasome have been reported, including boronic acids [133], epoxides [134], and lactones [135]. There are also several natural products that target the proteasome. One final class – the vinyl sulfones – modifies the catalytic threonine residue by acting as a Michael acceptor. The proteasome is composed of two copies of seven α and β subunits each. These subunits assemble into two types of seven subunits (α1–α7 and β1–β7). The 20S proteasome is formed by two rings of β subunits sandwiched on top and bottom by a ring of α subunits [136]. When capped by the 19S regulatory complex at both ends, the proteolytically active 26S proteasome is formed. In the eukaryotic proteasome, three of the seven β subunits, β1, β2 and β5 are responsible for the proteolytic action. Using vinyl sulfones as ABPP probes, Overkleeft and coworkers identified a broad-spectrum proteasome inhibitor that binds and covalently modifies the three active β subunits and enables monitoring of the proteasome both in vitro and in vivo [137] (Scheme 14).

Modifications of this approach have led to ABPP probes that are capable of selectively labeling the β1 and β5 subunits of the proteasome [138]. In further studies of the proteasome, bifunctional ABPP probes containing a vinyl sulfone linked to a photoreactive moiety allows photocrosslinking of subunits adjacent to proteasomal catalytic sites [139].

2.3.10 Protein Methyltransferase Substrates

Protein lysine methyltransferases (PKMTs) are a family of enzymes that transfer the activated methyl group from *S*-adenosyl-L-methionine (SAM) to specific lysine residues on various substrates. The PKMTs have been causally linked to various human diseases including cancer [140], Huntington's disease [141], and growth defects [142, 143]. The substrates of the PKMTs are typically histones [144–146], but there are several methyltransferases methylate non-histone substrates, such as the tumor suppressor p53 [147, 148], the estrogen receptor ERα [149], and the ATPase Reptin [150]. Given the importance of these enzymes in normal and

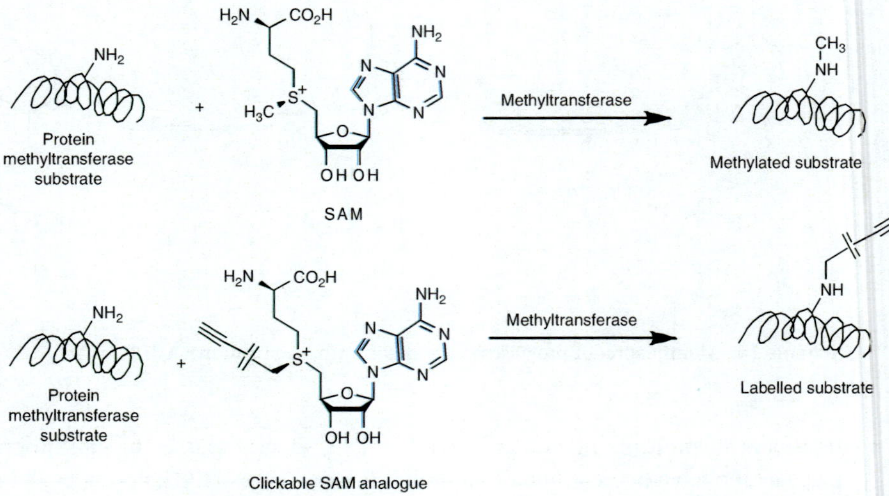

Fig. 10 SAM and "clickable" SAM analogs for labeling protein methyltransferase substrates

pathophysiology, extensive proteomic profiling of PKMT substrates will be important for indentifying non-histone and novel histone substrates of this enzyme class. To this effect, several groups have utilized a series of probes where the methyl group of SAM has been replaced by a "clickable" alkyne-containing side chain for labeling of PKMT substrates [151–153] (Fig. 10).

These groups have labeled purified histone H3 with these SAM/alkyne cofactors using different PKMTs. The lysine transferases that were found to accept the SAM/alkyne cofactor include the fungal Dim-5 [153] as well as the more complex human SETDB1 [151] and mixed-lineage leukemia PKMTs (MLL PKMTs), which are frequently complexed with additional subunits necessary for enzymatic activity [153]. This latter finding is of particular interest as MLL transferase subunits are frequently deregulated in cancer [154]. Although these PKMTs are able to accept the SAM/alkynes as cofactors, the enzymatic requirements for SAM/alkyne cofactor acceptance appear to be quite specific. For example, SAM/alkyne probe **1** is an active cofactor of SETDB1, but is inert towards SET7/9, SMYD2, PRDM8, PRDM10, and PRDM16 [151].

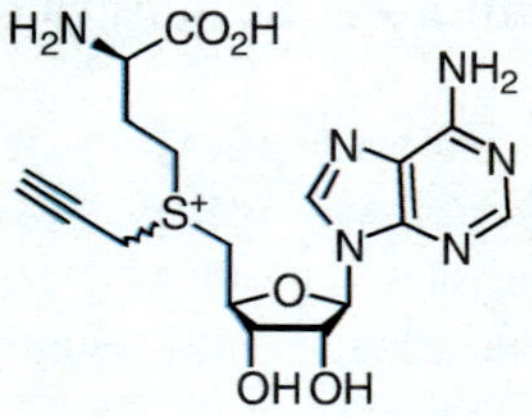

1

To address this problem, Islam and coworkers have engineered mutants of the PKMT G9a, a protein that is involved in methylating histone H3 at lysine 9 (H3K9), as well as other non-histone substrates [155, 156]. Human G9a has been implicated in oncogenesis through silencing tumor suppressors [157]. Modification of the catalytic channel of G9a by replacing conserved tyrosine and phenylalanine residues with alanine gave a mutant, Y1154A, which was effective in accepting SAM/alkyne probe **2** as a cofactor.

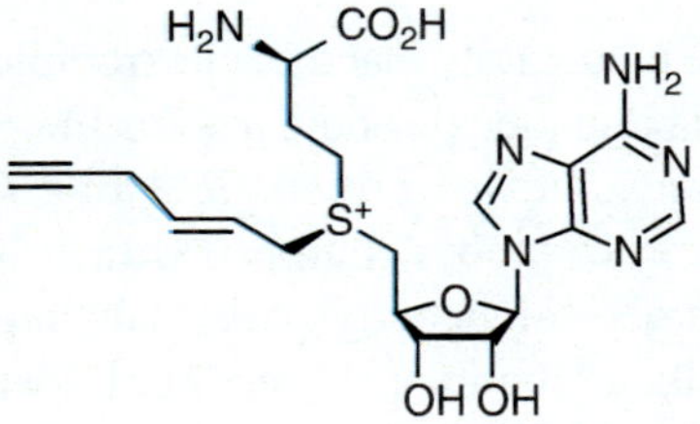

2

Using the mutant/cofactor **2** pair, the authors were able to exclusively modify the full-length human histone H3 in the presence of histone subunits H2A, H2B, and H4. Furthermore the MS data of the tryptic peptide revealed that the modification occurred at H3K9, indicating that the Y1154A mutant/cofactor **2** pair retains the same substrate specificity as the G9a/SAM pair. Further competition experiments revealed that the Y1154A/cofactor **2** pair was able to compete effectively for substrate labeling in the presence of equal amounts of G9a/SAM [152]. Later on, the same group successfully used the same approach to label the substrates of human protein arginine methyltransferase 1 via a double mutation of the wild-type enzyme [158]. While the "bump-hole" approach [159–161] as it applies to protein methyltransferase substrates has not yet been carried out in complex proteomes, the idea of engineering mutant/cofactor pairs for proteomic screening of methyltransferase substrates is an intriguing one, which may yield interesting results in the future.

2.3.11 Nondirected Probes

We have discussed ABPP probes that incorporate known binding groups specific to an enzyme class. The design of these directed probes requires previous knowledge of the enzymatic mechanism, active site structure, or ligand preference. Nondirected ABPP approaches involve the synthesis and screening of probes that bear a common electrophilic group and a variable binding group. The binding group is responsible for a variety of noncovalent interactions with the proteome, and is required for target selection. The reactive electrophilic group is present for the labeling of the selected protein targets. Conceptually, this approach is similar to target discovery via photoaffinity labeling, whereby a small molecule inhibitor with an unknown target is modified with a photoreactive label, e.g., a benzophenone moiety. The strength of nondirected ABPP strategies is that they can address enzymes and enzyme classes that lack affinity labels. To date, a number of electrophilic groups have been incorporated into nondirected ABPP probes. Some of the major classes of undirected probes will be discussed in this section.

Sulfonate Esters

Sulfonate esters are moderately reactive electrophiles that are easily modified at the sulfur atom, making them ideal candidates for library construction. To explore this untested electrophilic group in a proteomic setting, Cravatt and coworkers created a library of sulfonate ester probes equipped with either a biotin or a rhodamine tag. Specific protein targets of activity-based labeling were confirmed by heat- or pH-dependence in their labeling patterns [162]. Identification of the protein targets of the sulfonate ester ABPP library revealed the labeling of several mechanically distinct classes of enzymes, including dehydrogenases, glutathione *S*-transferases, sugar kinases, epoxide hydrolases, and transglutamases. None of the identified enzymes are targets of previously studied directed ABPP probes, and several of the enzymatic targets do not participate in covalent catalysis. Furthermore, sulfonate esters were shown to react with a large variety of weak nucleophiles, displaying labeling at tyrosine, aspartate, glutamate, and cysteine residues [36]. This promiscuous reactivity makes sulfonate ester-based ABPP probes a versatile chemotype for profiling a broad range of enzyme classes.

α-Chloroacetamide

The α-chloroacetamide group has features that are beneficial for undirected ABPP. Its small size does not bias binding elements towards a specific class of enzyme, and it possesses reactivity towards a broad variety of nucleophilic amino acid residues. A library of α-chloroacetamide-based probes were synthesized by Cravatt's group. The binding element in these probes was a dipeptide that was varied with small, large, hydrophobic, and charged side chains, and a biotin or rhodamine tag was appended as a reporter tag. Upon screening of eukaryotic proteomes with this library, many enzymes previously unaddressed by directed ABPP probes were uncovered. These included fatty acid synthase, hydroxypyruvate reductase, malic enzyme, and the nitrilase superfamily [163, 164]. In contrast to the sulfonate esters, α-chloroacetamides react preferentially with cysteine residues in the proteome.

Michael Acceptors

Of all the nondirected ABPP reactive groups, the Michael acceptor system is probably the most often used. This chemical moiety is found in many natural

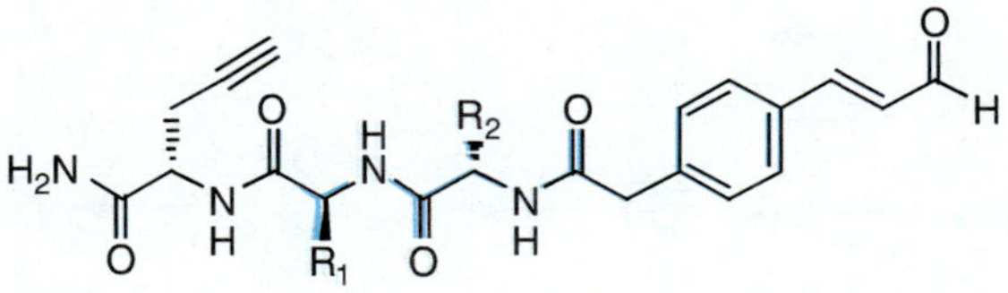

Fig. 11 Michael acceptor probes used by Pitscheider and Sieber [20]

products, often displaying mechanism-based activity via the 1,4-addition of cysteine residues, the preferred reacting partner of Michael acceptors in the proteome. Pitscheider and Sieber synthesized an undirected ABPP library based on cinnamic aldehyde as the reactive group [20]. The binding group consisted of a dipeptide moiety with varying side chains and an alkyne moiety for subsequent attachment of a reporter tag (Fig. 11).

Using this probe library, a virulence-associated secretory antigen, SsaA2 from methicillin-resistant *Staphylococcus aureus* (MRSA) was identified [20].

1,3-Cyclohexadiones

Reactive oxygen species can act as second messengers in signal transduction pathways that regulate cellular processes such as cell growth, differentiation, and migration [165–167]. It is likely that cysteine residues in proteins are the intracellular targets of oxidation, acting as redox sensors in the cell [168]. The first product in the reaction of cysteine and H_2O_2 is a sulfenic acid, which can then be reduced back to the thiol, or further oxidized to sulfinic or sulfonic acids [169]. Oxidation of cysteine to sulfenic acids in proteins has emerged as a biologically relevant post-translational modification of proteins; however, the identity of many of the modified proteins remains unknown. 1,3-Cyclohexadiones are known to undergo chemoselective reaction with sulfenic acids [170]. Leonard et al. have created a series of probes based on this chemotype that were used to selectively label both protein tyrosine phosphatase, as well as glyceraldehyde 3-phosphate dehydrogenase under oxidizing conditions, but not under reducing conditions [171] (Scheme 15).

When applied to whole HeLa cells, 193 proteins were identified, with diverse biological functions including signal transduction, protein synthesis, and chaperone-mediated protein folding. Of the proteins identified in this study, 56% were not previously reported to possess redox-active cysteines, and 93% were not reported to undergo sulfenic acid modification [172].

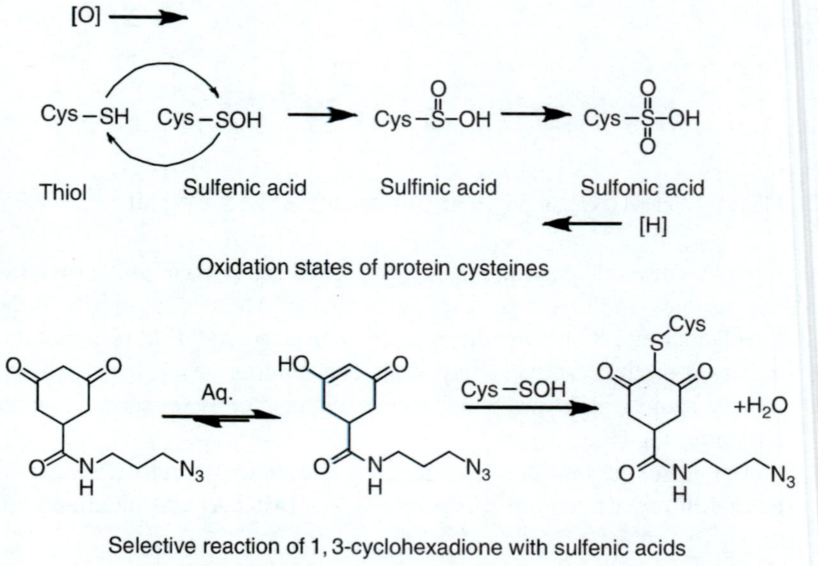

Scheme 15 Oxidation states of protein cysteines and labeling of 1,3-cyclohexadiones with sulfeninc acids

3 Biological Applications of ABPP

One of the main goals of proteomics is the assignment of proteins and pathways that are involved in physiological and pathological processes. The use of ABPP technologies can deliver information concerning the activity in native systems that leads to mechanistic insight concerning protein function. In the case of disease states, it is hoped that these functional assignments can assist the diagnosis and, eventually, treatment of human diseases. However, ABPP can be used for more than comparative proteomics as it pertains to pathological processes. In this section, we will provide an overview of some of the applications that ABPP offers to the field of proteomics.

3.1 *Enzyme Discovery and Classification*

Many gene products are uncharacterized enzymes that lack a specific class assignment. As discussed in Sect. 2.3, ABPP probes can be used to identify structurally disparate members of enzyme families based on their reactivity with mechanism-based inhibitors. To date, many uncharacterized enzymes have been classified upon their identification via ABPP. For example, the use of a fluorophosphonate probe by Jessani et al. led to the characterization of sialyl acetylesterase – expressed selectively in melanoma cell lines – as a member of the serine hydrolase superfamily [173]. This classification had been previously hindered by the lack of sequence homology between known serine hydrolases and sialyl acetylesterase [174]. The use of fluorophosphonate probes has been used to classify multiple unanticipated members of the serine hydrolase family from various eukaryotic proteomes. Another example of the use of ABPP probes to classify enzymes was reported by Li et al. who used a γ-secretase probe to categorize presenilin-1 as an aspartyl protease [175].

3.2 *Profiling Enzyme Activities in Disease States*

The elucidation of biochemical pathways in disease states is of fundamental importance in understanding and treating human illnesses. Comparative activity-based proteomics is a conceptually simple approach to determination of the identities of proteins involved in pathological disease states. In this approach, two types of cells – normal and diseased, or in the case of infectious disease, nonpathogenic and pathogenic bacteria – are treated with an appropriate ABPP probe. The labeling patterns of these two types of cells are then compared in an effort to gain insight into the molecular basis of the disease state. Comparative activity-based proteomics has been used to identify a number of proteins that play roles in various diseases.

3.2.1 Cancer

Cancerous cells are known to display a diverse set of metabolic changes from noncancer cells, including changes in glycolytic pathways [176, 177], the citric acid cycle [178], lipogenesis [179], and proteolysis [180]. Despite the advances in understanding cancer metabolism, the current picture is far from complete. ABPP can be used to functionally characterize deregulated enzymatic activities in cancer. Proteases play crucial roles in cancer because they can contribute to the degradation of connective tissue and dissolution of epithelial and endothelial basement membranes. This can manifest itself in the remodeling processes that occurs during tumor invasion and metastasis [1, 181]. As outlined in Sect. 2.3.1, fluorophosphonates can be used to target the serine hydrolase superfamily. Using these probes, Jessani et al. have been able to discover several unregulated enzymes in cancer [44, 60, 61, 182]. Other protease families can be profiled using ABPP probes. For example, using epoxide probes for cysteine proteases, Joyce et al. found that cathepsin activity is higher in angiogenic vasculature and at the invasive fronts of carcinomas. They also found that cysteine cathepsins are increased in human papilloma virus-induced cervical carcinomas [183]. Profiling metalloproteinase activities with photoactive, hydroxamate activity-based probes

(Sect. 2.3.3) has uncovered a membrane-associated glycoprotein that has increased activity in aggressive human melanoma lines as compared to less-aggressive cell lines [85].

3.2.2 Infectious Diseases

Pathogenic bacterial infections by multidrug-resistant bacteria are becoming more and more untreatable using currently available antibiotics, raising concerns of a return to a "pre-antibiotic" era [184]. Currently available antibacterial agents address only a limited number of targets, such as DNA replication and protein and cell wall synthesis. It is therefore a pressing concern in the medical community to identify novel targets in resistant bacteria for functional characterization. Heal and Tate [185] cover the topic of using ABPP as it applies to microbial pathogenesis and potential novel antibacterial targets.

3.3 Imaging of Enzyme Activities

The ability of ABPP to provide information on native protein systems can be adapted to analyze enzyme function in living cells and organisms via fluorescence. Bogyo and coworkers have developed fluorescently quenched activity-based probes that are capable of visualizing cathepsin activity in living cells [186]. Cathepsins have been strongly implicated in initiating and promoting tumor formation, growth, invasiveness, and metastasis [180, 187]. Increased expression and activity of cathepsin B and cathepsin L is found in a variety of malignancies and are strong negative prognostic indicators in individuals with cancer [188, 189]. As cathepsins are cysteine proteases, a cell-permeable probe based on an acyloxymethylketone (AOMK) reactive group (Sect. 2.3.2) containing a fluorescent group on the probe and a quenching group on the leaving group was synthesized. Upon covalent attachment to a protease, the quenching group is liberated, effectively fluorescently labeling the modified enzyme. These probes were then used for noninvasive, whole body imaging of live mice bearing grafted tumors. Probes without the AOMK reactive group were cleared from the mice, resulting in a loss of background fluorescence within 5 h. Mice treated with the AOMK-bearing probes displayed elevated fluorescence at the tumor sites compared to the background. Furthermore, the permanent nature of the protein modification by the AOMK probes allowed the direct monitoring and identification of specifically labeled proteins by SDS-PAGE analysis of tissue homogenates collected from animals treated with the probe. In another elegant example of in vivo enzyme imaging, a recent report describes the labeling of the transcriptional factor LasR. The LasR protein is a quorum-sensing signal receptor that plays a pivotal role in the activation of many virulence genes in *Pseudomonas aeruginosa*. The induction of LasR is carried out by its cognate autoinducer, 3-oxo-dodecanoyl homoserine lactone (3-oxo-C12-HSL), which is

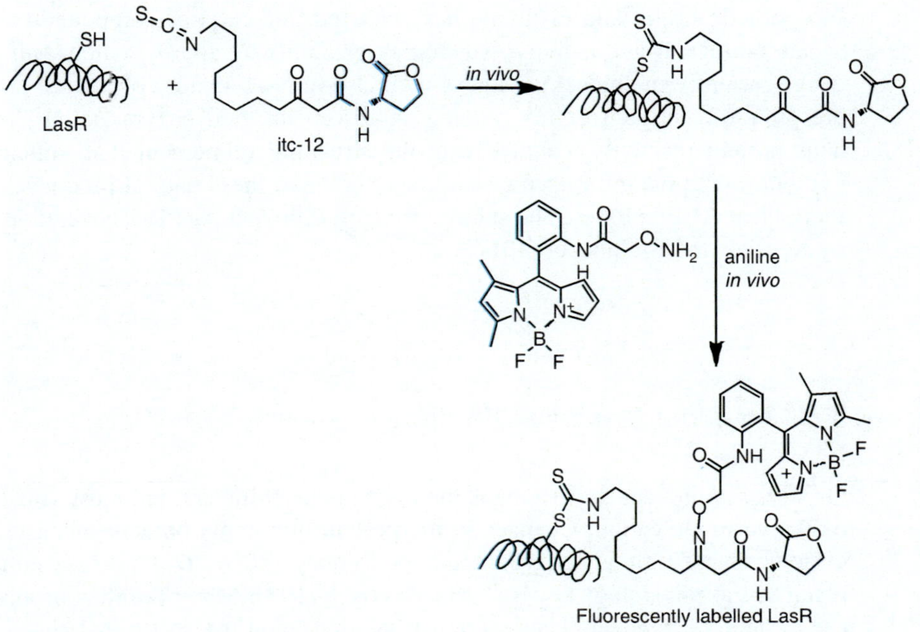

Scheme 16 Imaging of LasR with Itc-12 probe via bio-orthogonal imine formation

the primary quorum sensing signal in *P. aeruginosa*. Analysis of the X-ray structure of LasR co-crystallized with 3-oxo-C12-HSL shows that there is a cysteine residue (cysteine 79) in close proximity to the end of the alkyl chain of 3-oxo-C12-HSL [131]. Incorporating an isothiocyanate group to the terminus of this alkyl chain gives the covalent inhibitor itc-12 [34]. Using itc-12 to label LasR, Rayo et al. took advantage of the itc-12 structure to perform a bio-orthogonal reaction between BODIPY-ONH$_2$ and the ketone moiety of the bound inhibitor [190]. With aniline as a catalyst, this reaction proceeds in vitro to completion in 12 h at pH 6.6 and 8 °C. The in vivo reaction between itc-12 and LasR followed by the subsequent BODIPY-ONH$_2$ fluorescent conjugation in *P. aeruginosa* proceeds with very high selectivity, fluorescently labeling only LasR, as determined by in-gel fluorescent imaging and MS/MS analysis (Scheme 16).

Imaging of LasR in live *P. aeruginosa* cells via this method reveals that LasR is not distributed evenly throughout the cytoplasmic membrane, but is rather concentrated at the poles of the *P. aeruginosa* cells [190].

3.4 Natural Product Target Elucidation

Natural products have found use in human medicine for thousands of years, and their impact on human health and well-being cannot be overstated. Their biosynthetic origin and evolution has endowed this class of compounds with an innate biological interactivity, which may explain why approximately half of the drugs currently in clinical use are of natural origin. A number of natural products are protein-reactive, forming covalent adducts with their enzymatic targets [19]. This protein reactivity, coupled with the structural refinement that millennia of evolution with natural systems have imparted upon these natural products, makes them ideal ABPP probes. The subject of natural product ABPP is covered in detail by Krysiak and Breinbauer [191].

3.5 Cysteine Reactivity Profiling

Of all the amino acids, cysteine is the most nucleophilic and the most sensitive to oxidative modification. Changes in the protein microenvironment of the thiol in cysteine can result in the formation of thiolate anions that possess enhanced reactivity at physiological pH (hyper-reactive) [192]. Many families of enzymes use cysteine-dependent chemical transformations, and the post-translational modification of cysteine as disulfides, glutathiones, and sulfenic, sulfinic, and sulfonic acids gives cysteines the ability to serve as a regulatory switch in proteins [169]. Evidence suggests that nucleophilic cysteines may possess a variety of functions outside of catalysis; however, measurement of hyper-reactive cysteine pK_a requires purified protein combined with detailed kinetic and mutagenic experiments, precluding measurement on a proteome-wide scale [193, 194]. To address this, Weerapana et al. developed a strategy to globally characterize cysteine functionality in proteomes with isotopically labeled, ABPP probes [195]. In this approach, named isotopic tandem orthogonal proteolysis–activity-based protein profiling (isoTOP-ABPP), a cysteine-reactive iodoacetamide ABPP probe is applied in two concentrations (high and low) to aliquots of a proteome. After protein labeling, the bound probes are tagged with a biotin attached to an isotopically labeled linker (heavy and light) containing a TEV protease site. After mixing and enrichment via avidin beads, followed by tandem proteolytic on-bead digestions with trypsin and TEV protease, respectively, a mix of isotopically labeled peptides is obtained. These are then analyzed by LC–high resolution MS, and the extent of labeling determined for the heavy and light variants. When accounting for concentration differences, the ratio of labeling between heavy and light variants $R_{[high][low]}$ in the two experiments gives a measurement of the extent of cysteine labeling (Fig. 12).

Using the isoTOP-ABPP method, the soluble proteome of the breast cancer cell line MCF7 was interrogated and, of the 800 probe-labeled cysteines, 10% were found to be hyper-reactive ($R_{10:1} < 2$). A database search further revealed that hyper-reactive cysteines were over-represented in functional residues (active site

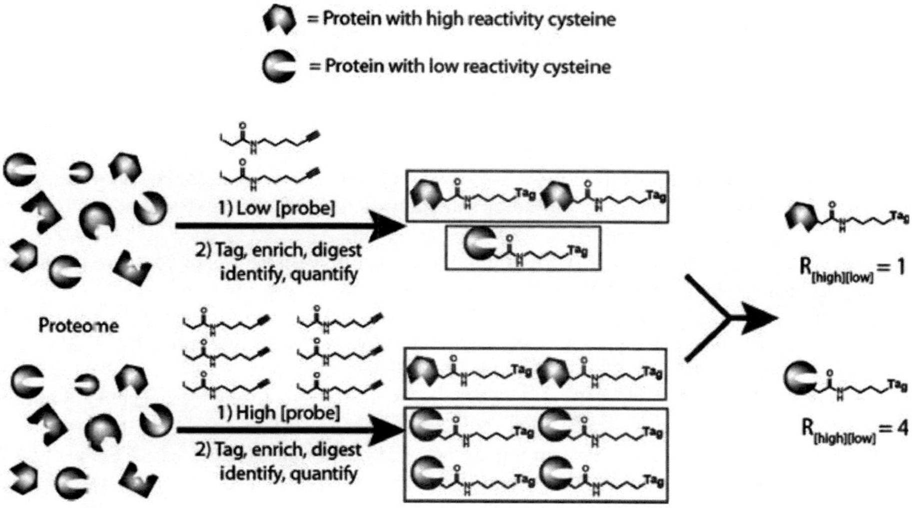

Fig. 12 isoTOP-ABPP for cysteine reactivity profiling. $R_{[\mathrm{high}][\mathrm{low}]}$ is the ratio of labeling between heavy and light variants

nucleophiles, redox-active cysteines, sites of post-translational modification) and tended to be conserved across eukaryotic evolution.

4 Conclusions and Outlook

In this introductory chapter, we have aimed to expose the reader to the many facets of ABPP. Over the past years, ABPP has become a standard technique for the rapid, sensitive, and selective identification of enzyme activities in complex proteomes. ABPP is a highly interdisciplinary endeavor, combining organic synthesis, analytical chemistry, mechanistic enzymology, and cell biology, and this interdisciplinary character allows it to address biological questions that cannot be addressed by one discipline alone. Currently, a number of ABPP chemotypes exist that are successful for the functional characterization of many diseases such as cancer, bacterial infection, and metabolic disorders. This, in combination applications for competitive inhibitor screening and molecular imaging, renders ABPP a key technology for chemical biology and biomedical research. A major task for future ABPP studies will be the expansion of available chemotypes that are capable of profiling a complete set of physiological and pharmacological enzyme classes, including those that have not yet been addressed, such as ion channels and structural proteins.

Acknowledgments M. Nodwell thanks the Alexander von Humboldt foundation for a postdoctoral fellowship. S. Sieber is supported by the Deutsche Forschungsgemeinschaft (Emmy Noether), SFB749, FOR1406, an ERC starting grant, and the Center for Integrated Protein Science Munich CIPSM.

References

1. Hanahan D, Weinberg RA (2000) The hallmarks of cancer. Cell 100:57–70
2. Vogelstein B, Kinzler KW (2004) Cancer genes and the pathways they control. Nat Med 10:789–799
3. Kobe B, Kemp BE (1999) Active site-directed protein regulation. Nature 402:373–376
4. Walsh CT (2006) Posttranslational modification of proteins: expanding nature's inventory. Roberts, Englewood, CO
5. Gygi SP, Rist B, Gerber SA et al (1999) Quantitative analysis of complex protein mixtures using isotope-coded affinity tags. Nat Biotechnol 17:994–999
6. Washburn MP, Wolters D, Yates JR 3rd (2001) Large-scale analysis of the yeast proteome by multidimensional protein identification technology. Nat Biotechnol 19:242–247
7. Zhu H, Bilgin M, Snyder M (2003) Proteomics. Annu Rev Biochem 72:783–812
8. Ito T, Ota K, Kubota H et al (2002) Roles for the two-hybrid system in exploration of the yeast protein interactome. Mol Cell Proteomics 1:561–566
9. MacBeath G (2002) Protein microarrays and proteomics. Nat Genet 32(Suppl):526–532
10. Evans MJ, Cravatt BF (2006) Mechanism-based profiling of enzyme families. Chem Rev 106:3279–3301
11. Heal WP, Dang TH, Tate EW (2011) Activity-based probes: discovering new biology and new drug targets. Chem Soc Rev 40:246–257
12. Cravatt BF, Wright AT, Kozarich JW (2008) Activity-based protein profiling: from enzyme chemistry to proteomic chemistry. Annu Rev Biochem 77:383–414
13. Bottcher T, Pitscheider M, Sieber SA (2010) Natural products and their biological targets: proteomic and metabolomic labeling strategies. Angew Chem Int Ed Engl 49:2680–2698
14. Barglow KT, Cravatt BF (2007) Activity-based protein profiling for the functional annotation of enzymes. Nat Methods 4:822–827
15. Dennehy MK, Richards KA, Wernke GR et al (2006) Cytosolic and nuclear protein targets of thiol-reactive electrophiles. Chem Res Toxicol 19:20–29
16. Scaloni A, Ferranti P, De Simone G et al (1999) Probing the reactivity of nucleophile residues in human 2,3-diphosphoglycerate/deoxy-hemoglobin complex by aspecific chemical modifications. FEBS Lett 452:190–194
17. Slaughter DE, Hanzlik RP (1991) Identification of epoxide- and quinone-derived bromobenzene adducts to protein sulfur nucleophiles. Chem Res Toxicol 4:349–359
18. Rando RR (1977) Mechanism-based irreversible enzyme inhibitors. Methods Enzymol 46:28–41
19. Drahl C, Cravatt BF, Sorensen EJ (2005) Protein-reactive natural products. Angew Chem Int Ed Engl 44:5788–5809
20. Pitscheider M, Sieber SA (2009) Cinnamic aldehyde derived probes for the active site labeling of pathogenesis associated enzymes. Chem Commun 2009:3741–3743
21. Weerapana E, Simon GM, Cravatt BF (2008) Disparate proteome reactivity profiles of carbon electrophiles. Nat Chem Biol 4:405–407
22. Evans MJ, Saghatelian A, Sorensen EJ et al (2005) Target discovery in small-molecule cell-based screens by in situ proteome reactivity profiling. Nat Biotechnol 23:1303–1307
23. Robinette D, Neamati N, Tomer KB et al (2006) Photoaffinity labeling combined with mass spectrometric approaches as a tool for structural proteomics. Expert Rev Proteomics 3:399–408
24. Tanaka Y, Bond MR, Kohler JJ (2008) Photocrosslinkers illuminate interactions in living cells. Mol Biosyst 4:473–480
25. Bissantz C, Kuhn B, Stahl M (2010) A medicinal chemist's guide to molecular interactions. J Med Chem 53:5061–5084
26. Rostovtsev VV, Green LG, Fokin VV et al (2002) A stepwise Huisgen cycloaddition process: copper(I)-catalyzed regioselective "ligation" of azides and terminal alkynes. Angew Chem Int Ed Engl 41:2596–2599

27. Meldal M, Tornoe CW (2008) Cu-catalyzed azide-alkyne cycloaddition. Chem Rev 108:2952–3015
28. Speers AE, Adam GC, Cravatt BF (2003) Activity-based protein profiling in vivo using a copper(I)-catalyzed azide-alkyne [3+2] cycloaddition. J Am Chem Soc 125:4686–4687
29. Saxon E, Bertozzi CR (2000) Cell surface engineering by a modified Staudinger reaction. Science 287:2007–2010
30. Kohn M, Breinbauer R (2004) The Staudinger ligation-a gift to chemical biology. Angew Chem Int Ed Engl 43:3106–3116
31. Agard NJ, Prescher JA, Bertozzi CR (2004) A strain-promoted [3+2] azide-alkyne cycloaddition for covalent modification of biomolecules in living systems. J Am Chem Soc 126:15046–15047
32. Beatty KE, Fisk JD, Smart BP et al (2010) Live-cell imaging of cellular proteins by a strain-promoted azide-alkyne cycloaddition. Chembiochem 11:2092–2095
33. Codelli JA, Baskin JM, Agard NJ et al (2008) Second-generation difluorinated cyclooctynes for copper-free click chemistry. J Am Chem Soc 130:11486–11493
34. Amara N, Mashiach R, Amar D et al (2009) Covalent inhibition of bacterial quorum sensing. J Am Chem Soc 131:10610–10619
35. Liu S, Zhou B, Yang H et al (2008) Aryl vinyl sulfonates and sulfones as active site-directed and mechanism-based probes for protein tyrosine phosphatases. J Am Chem Soc 130:8251–8260
36. Adam GC, Sorensen EJ, Cravatt BF (2002) Proteomic profiling of mechanistically distinct enzyme classes using a common chemotype. Nat Biotechnol 20:805–809
37. Patricelli MP, Giang DK, Stamp LM et al (2001) Direct visualization of serine hydrolase activities in complex proteomes using fluorescent active site-directed probes. Proteomics 1:1067–1071
38. Liu Y, Patricelli MP, Cravatt BF (1999) Activity-based protein profiling: the serine hydrolases. Proc Natl Acad Sci USA 96:14694–14699
39. Kidd D, Liu Y, Cravatt BF (2001) Profiling serine hydrolase activities in complex proteomes. Biochemistry 40:4005–4015
40. Santoni V, Molloy M, Rabilloud T (2000) Membrane proteins and proteomics: un amour impossible? Electrophoresis 21:1054–1070
41. Corthals GL, Wasinger VC, Hochstrasser DF et al (2000) The dynamic range of protein expression: a challenge for proteomic research. Electrophoresis 21:1104–1115
42. Bottcher T, Sieber SA (2010) Showdomycin as a versatile chemical tool for the detection of pathogenesis-associated enzymes in bacteria. J Am Chem Soc 132:6964–6972
43. Tully SE, Cravatt BF (2010) Activity-based probes that target functional subclasses of phospholipases in proteomes. J Am Chem Soc 132:3264–3265
44. Jessani N, Niessen S, Wei BQ et al (2005) A streamlined platform for high-content functional proteomics of primary human specimens. Nat Methods 2:691–697
45. Liu H, Sadygov RG, Yates JR 3rd (2004) A model for random sampling and estimation of relative protein abundance in shotgun proteomics. Anal Chem 76:4193–4201
46. Old WM, Meyer-Arendt K, Aveline-Wolf L et al (2005) Comparison of label-free methods for quantifying human proteins by shotgun proteomics. Mol Cell Proteomics 4:1487–1502
47. Adam GC, Burbaum J, Kozarich JW et al (2004) Mapping enzyme active sites in complex proteomes. J Am Chem Soc 126:1363–1368
48. Okerberg ES, Wu J, Zhang B et al (2005) High-resolution functional proteomics by active-site peptide profiling. Proc Natl Acad Sci USA 102:4996–5001
49. Speers AE, Cravatt BF (2005) A tandem orthogonal proteolysis strategy for high-content chemical proteomics. J Am Chem Soc 127:10018–10019
50. Sieber SA, Mondala TS, Head SR et al (2004) Microarray platform for profiling enzyme activities in complex proteomes. J Am Chem Soc 126:15640–15641
51. Kodadek T (2001) Protein microarrays: prospects and problems. Chem Biol 8:105–115

52. Winssinger N, Ficarro S, Schultz PG et al (2002) Profiling protein function with small molecule microarrays. Proc Natl Acad Sci USA 99:11139–11144
53. Clark JD, Schievella AR, Nalefski EA et al (1995) Cytosolic phospholipase A2. J Lipid Mediat Cell Signal 12:83–117
54. Mignatti P, Rifkin DB (1996) Plasminogen activators and angiogenesis. Curr Top Microbiol Immunol 213(Pt 1):33–50
55. DeClerck YA, Imren S, Montgomery AM et al (1997) Proteases and protease inhibitors in tumor progression. Adv Exp Med Biol 425:89–97
56. Gorrell MD (2005) Dipeptidyl peptidase IV and related enzymes in cell biology and liver disorders. Clin Sci (Lond) 108:277–292
57. Walsh CT (1979) Enzymatic reaction mechanisms. W.H. Freeman, New York
58. Powers JC, Asgian JL, Ekici OD et al (2002) Irreversible inhibitors of serine, cysteine, and threonine proteases. Chem Rev 102:4639–4750
59. Bouma BN, Miles LA, Beretta G et al (1980) Human plasma prekallikrein. Studies of its activation by activated factor XII and of its inactivation by diisopropyl phosphofluoridate. Biochemistry 19:1151–1160
60. Jessani N, Humphrey M, McDonald WH et al (2004) Carcinoma and stromal enzyme activity profiles associated with breast tumor growth *in vivo*. Proc Natl Acad Sci USA 101:13756–13761
61. Jessani N, Liu Y, Humphrey M et al (2002) Enzyme activity profiles of the secreted and membrane proteome that depict cancer cell invasiveness. Proc Natl Acad Sci USA 99:10335–10340
62. Mahrus S, Craik CS (2005) Selective chemical functional probes of granzymes A and B reveal granzyme B is a major effector of natural killer cell-mediated lysis of target cells. Chem Biol 12:567–577
63. Gelb BD, Shi GP, Chapman HA et al (1996) Pycnodysostosis, a lysosomal disease caused by cathepsin K deficiency. Science 273:1236–1238
64. Sloane BF, Yan S, Podgorski I et al (2005) Cathepsin B and tumor proteolysis: contribution of the tumor microenvironment. Semin Cancer Biol 15:149–157
65. Yan S, Sameni M, Sloane BF (1998) Cathepsin B and human tumor progression. Biol Chem 379:113–123
66. Shenai BR, Sijwali PS, Singh A et al (2000) Characterization of native and recombinant falcipain-2, a principal trophozoite cysteine protease and essential hemoglobinase of *Plasmodium falciparum*. J Biol Chem 275:29000–29010
67. Iwata Y, Mort JS, Tateishi H et al (1997) Macrophage cathepsin L, a factor in the erosion of subchondral bone in rheumatoid arthritis. Arthritis Rheum 40:499–509
68. Barrett AJ, Kembhavi AA, Brown MA et al (1982) L-trans-Epoxysuccinyl-leucylamido(4-guanidino)butane (E-64) and its analogues as inhibitors of cysteine proteinases including cathepsins B, H and L. Biochem J 201:189–198
69. Barrett AJ, Kembhavi AA, Hanada K (1981) E-64 [L-trans-epoxysuccinyl-leucyl-amido(4-guanidino)butane] and related epoxides as inhibitors of cysteine proteinases. Acta Biol Med Ger 40:1513–1517
70. Palmer JT, Rasnick D, Klaus JL et al (1995) Vinyl sulfones as mechanism-based cysteine protease inhibitors. J Med Chem 38:3193–3196
71. Shaw E (1994) Peptidyl diazomethanes as inhibitors of cysteine and serine proteinases. Methods Enzymol 244:649–656
72. Shaw E, Angliker H, Rauber P et al (1986) Peptidyl fluoromethyl ketones as thiol protease inhibitors. Biomed Biochim Acta 45:1397–1403
73. Pliura DH, Bonaventura BJ, Smith RA et al (1992) Comparative behaviour of calpain and cathepsin B toward peptidyl acyloxymethyl ketones, sulphonium methyl ketones and other potential inhibitors of cysteine proteinases. Biochem J 288(Pt 3):759–762
74. Fonovic M, Bogyo M (2007) Activity based probes for proteases: applications to biomarker discovery, molecular imaging and drug screening. Curr Pharm Des 13:253–261

75. Brady KD, Giegel DA, Grinnell C et al (1999) A catalytic mechanism for caspase-1 and for bimodal inhibition of caspase-1 by activated aspartic ketones. Bioorg Med Chem 7:621–631
76. Dai Y, Hedstrom L, Abeles RH (2000) Inactivation of cysteine proteases by (acyloxy)methyl ketones using S′-P′ interactions. Biochemistry 39:6498–6502
77. Kato D, Boatright KM, Berger AB et al (2005) Activity-based probes that target diverse cysteine protease families. Nat Chem Biol 1:33–38
78. Mitic N, Smith SJ, Neves A et al (2006) The catalytic mechanisms of binuclear metallohydrolases. Chem Rev 106:3338–3363
79. Whittaker M, Floyd CD, Brown P et al (1999) Design and therapeutic application of matrix metalloproteinase inhibitors. Chem Rev 99:2735–2776
80. Skiles JW, Gonnella NC, Jeng AY (2001) The design, structure, and therapeutic application of matrix metalloproteinase inhibitors. Curr Med Chem 8:425–474
81. Hooper NM, Turner AJ (2002) The search for alpha-secretase and its potential as a therapeutic approach to Alzheimer s disease. Curr Med Chem 9:1107–1119
82. Vihinen P, Kahari VM (2002) Matrix metalloproteinases in cancer: prognostic markers and therapeutic targets. Int J Cancer 99:157–166
83. Sierevogel MJ, Pasterkamp G, de Kleijn DP et al (2003) Matrix metalloproteinases: a therapeutic target in cardiovascular disease. Curr Pharm Des 9:1033–1040
84. Sieber SA, Niessen S, Hoover HS et al (2006) Proteomic profiling of metalloprotease activities with cocktails of active-site probes. Nat Chem Biol 2:274–281
85. Saghatelian A, Jessani N, Joseph A et al (2004) Activity-based probes for the proteomic profiling of metalloproteases. Proc Natl Acad Sci USA 101:10000–10005
86. Chan EW, Chattopadhaya S, Panicker RC et al (2004) Developing photoactive affinity probes for proteomic profiling: hydroxamate-based probes for metalloproteases. J Am Chem Soc 126:14435–14446
87. Minucci S, Pelicci PG (2006) Histone deacetylase inhibitors and the promise of epigenetic (and more) treatments for cancer. Nat Rev Cancer 6:38–51
88. Bolden JE, Peart MJ, Johnstone RW (2006) Anticancer activities of histone deacetylase inhibitors. Nat Rev Drug Discov 5:769–784
89. Marks PA, Breslow R (2007) Dimethyl sulfoxide to vorinostat: development of this histone deacetylase inhibitor as an anticancer drug. Nat Biotechnol 25:84–90
90. Salisbury CM, Cravatt BF (2007) Activity-based probes for proteomic profiling of histone deacetylase complexes. Proc Natl Acad Sci USA 104:1171–1176
91. Overkleeft HS et al. (2012) Photoaffinity labeling in activity-based protein profiling. Topics in Current Chemistry. Springer, Berlin, Heidelberg. doi: 10.1007/128_2011_XXX
92. Manning G, Whyte DB, Martinez R et al (2002) The protein kinase complement of the human genome. Science 298:1912–1934
93. Elphick LM, Lee SE, Gouverneur V et al (2007) Using chemical genetics and ATP analogs to dissect protein kinase function. ACS Chem Biol 2:299–314
94. Krause DS, Van Etten RA (2005) Tyrosine kinases as targets for cancer therapy. N Engl J Med 353:172–187
95. Wymann MP, Marone R (2005) Phosphoinositide 3-kinase in disease: timing, location, and scaffolding. Curr Opin Cell Biol 17:141–149
96. Noble ME, Endicott JA, Johnson LN (2004) Protein kinase inhibitors: insights into drug design from structure. Science 303:1800–1805
97. Fedorov O, Marsden B, Pogacic V et al (2007) A systematic interaction map of validated kinase inhibitors with Ser/Thr kinases. Proc Natl Acad Sci USA 104:20523–20528
98. Fedorov O, Sundstrom M, Marsden B et al (2007) Insights for the development of specific kinase inhibitors by targeted structural genomics. Drug Discov Today 12:365–372
99. Hanks SK, Hunter T (1995) Protein kinases 6. The eukaryotic protein kinase superfamily: kinase (catalytic) domain structure and classification. FASEB J 9:576–596

100. Zheng J, Knighton DR, ten Eyck LF et al (1993) Crystal structure of the catalytic subunit of cAMP-dependent protein kinase complexed with MgATP and peptide inhibitor. Biochemistry 32:2154–2161
101. Patricelli MP, Szardenings AK, Liyanage M et al (2007) Functional interrogation of the kinome using nucleotide acyl phosphates. Biochemistry 46:350–358
102. Arcaro A, Wymann MP (1993) Wortmannin is a potent phosphatidylinositol 3-kinase inhibitor: the role of phosphatidylinositol 3,4,5-trisphosphate in neutrophil responses. Biochem J 296(Pt 2):297–301
103. Yano H, Nakanishi S, Kimura K et al (1993) Inhibition of histamine secretion by wortmannin through the blockade of phosphatidylinositol 3-kinase in RBL-2H3 cells. J Biol Chem 268:25846–25856
104. Breinbauer R et al. (2012) Activity Based Protein Profiling for Natural Product Target Discovery. Topics in Current Chemistry. Springer, Berlin, Heidelberg. doi: 10.1007/128_2011_XXX
105. Gupta V, Ogawa AK, Du X et al (1997) A model for binding of structurally diverse natural product inhibitors of protein phosphatases PP1 and PP2A. J Med Chem 40:3199–3206
106. Zhou L, Yu H, Chen K (2002) Relationship between microcystin in drinking water and colorectal cancer. Biomed Environ Sci 15:166–171
107. Goldberg J, Huang HB, Kwon YG et al (1995) Three-dimensional structure of the catalytic subunit of protein serine/threonine phosphatase-1. Nature 376:745–753
108. Shreder KR, Liu Y, Nomanhboy T et al (2004) Design and synthesis of AX7574: a microcystin-derived, fluorescent probe for serine/threonine phosphatases. Bioconjug Chem 15:790–798
109. Lo LC, Pang TL, Kuo CH et al (2002) Design and synthesis of class-selective activity probes for protein tyrosine phosphatases. J Proteome Res 1:35–40
110. Kumar S, Zhou B, Liang F et al (2004) Activity-based probes for protein tyrosine phosphatases. Proc Natl Acad Sci USA 101:7943–7948
111. Zechel DL, Withers SG (2000) Glycosidase mechanisms: anatomy of a finely tuned catalyst. Acc Chem Res 33:11–18
112. Tsai CS, Li YK, Lo LC (2002) Design and synthesis of activity probes for glycosidases. Org Lett 4:3607–3610
113. Wicki J, Rose DR, Withers SG (2002) Trapping covalent intermediates on beta-glycosidases. Methods Enzymol 354:84–105
114. Vocadlo DJ, Bertozzi CR (2004) A strategy for functional proteomic analysis of glycosidase activity from cell lysates. Angew Chem Int Ed Engl 43:5338–5342
115. Vocadlo DJ, Hang HC, Kim EJ et al (2003) A chemical approach for identifying O-GlcNAc-modified proteins in cells. Proc Natl Acad Sci USA 100:9116–9121
116. Denisov IG, Makris TM, Sligar SG et al (2005) Structure and chemistry of cytochrome P450. Chem Rev 105:2253–2277
117. Guengerich FP, Wu ZL, Bartleson CJ (2005) Function of human cytochrome P450s: characterization of the orphans. Biochem Biophys Res Commun 338:465–469
118. Hughes AL, Powell DW, Bard M et al (2007) Dap1/PGRMC1 binds and regulates cytochrome P450 enzymes. Cell Metab 5:143–149
119. Aguiar M, Masse R, Gibbs BF (2005) Regulation of cytochrome P450 by posttranslational modification. Drug Metab Rev 37:379–404
120. Wright AT, Cravatt BF (2007) Chemical proteomic probes for profiling cytochrome p450 activities and drug interactions in vivo. Chem Biol 14:1043–1051
121. Wright AT, Song JD, Cravatt BF (2009) A suite of activity-based probes for human cytochrome P450 enzymes. J Am Chem Soc 131:10692–10700
122. Jones JE, Causey CP, Knuckley B et al (2009) Protein arginine deiminase 4 (PAD4): current understanding and future therapeutic potential. Curr Opin Drug Discov Devel 12:616–627
123. Vossenaar ER, van Venrooij WJ (2004) Citrullinated proteins: sparks that may ignite the fire in rheumatoid arthritis. Arthritis Res Ther 6:107–111

124. Luo Y, Arita K, Bhatia M et al (2006) Inhibitors and inactivators of protein arginine deiminase 4: functional and structural characterization. Biochemistry 45:11727–11736
125. Luo Y, Knuckley B, Lee YH et al (2006) A fluoroacetamidine-based inactivator of protein arginine deiminase 4: design, synthesis, and *in vitro* and *in vivo* evaluation. J Am Chem Soc 128:1092–1093
126. Stone EM, Schaller TH, Bianchi H et al (2005) Inactivation of two diverse enzymes in the amidinotransferase superfamily by 2-chloroacetamidine: dimethylargininase and peptidylarginine deiminase. Biochemistry 44:13744–13752
127. Slack JL, Causey CP, Luo Y et al (2011) Development and use of clickable activity based protein profiling agents for protein arginine deiminase 4. ACS Chem Biol 6:466–476
128. Luo Y, Knuckley B, Bhatia M et al (2006) Activity-based protein profiling reagents for protein arginine deiminase 4 (PAD4): synthesis and *in vitro* evaluation of a fluorescently labeled probe. J Am Chem Soc 128:14468–14469
129. Coux O, Tanaka K, Goldberg AL (1996) Structure and functions of the 20S and 26S proteasomes. Annu Rev Biochem 65:801–847
130. Bogyo M, McMaster JS, Gaczynska M et al (1997) Covalent modification of the active site threonine of proteasomal beta subunits and the Escherichia coli homolog HslV by a new class of inhibitors. Proc Natl Acad Sci USA 94:6629–6634
131. Bochtler M, Ditzel L, Groll M et al (1999) The proteasome. Annu Rev Biophys Biomol Struct 28:295–317
132. Jackson G, Einsele H, Moreau P et al (2005) Bortezomib, a novel proteasome inhibitor, in the treatment of hematologic malignancies. Cancer Treat Rev 31:591–602
133. Richardson PG, Mitsiades C, Hideshima T et al (2006) Bortezomib: proteasome inhibition as an effective anticancer therapy. Annu Rev Med 57:33–47
134. Meng L, Kwok BH, Sin N et al (1999) Eponemycin exerts its antitumor effect through the inhibition of proteasome function. Cancer Res 59:2798–2801
135. Chauhan D, Catley L, Li G et al (2005) A novel orally active proteasome inhibitor induces apoptosis in multiple myeloma cells with mechanisms distinct from Bortezomib. Cancer Cell 8:407–419
136. Baumeister W, Walz J, Zuhl F et al (1998) The proteasome: paradigm of a self-compartmentalizing protease. Cell 92:367–380
137. Verdoes M, Florea BI, Menendez-Benito V et al (2006) A fluorescent broad-spectrum proteasome inhibitor for labeling proteasomes *in vitro* and *in vivo*. Chem Biol 13:1217–1226
138. Verdoes M, Willems LI, van der Linden WA et al (2010) A panel of subunit-selective activity-based proteasome probes. Org Biomol Chem 8:2719–2727
139. Geurink PP, Florea BI, Van der Marel GA et al (2010) Probing the proteasome cavity in three steps: bio-orthogonal photo-reactive suicide substrates. Chem Commun (Camb) 46:9052–9054
140. Spannhoff A, Sippl W, Jung M (2009) Cancer treatment of the future: inhibitors of histone methyltransferases. Int J Biochem Cell Biol 41:4–11
141. Ryu H, Lee J, Hagerty SW et al (2006) ESET/SETDB1 gene expression and histone H3 (K9) trimethylation in Huntington's disease. Proc Natl Acad Sci USA 103:19176–19181
142. Berdasco M, Ropero S, Setien F et al (2009) Epigenetic inactivation of the Sotos overgrowth syndrome gene histone methyltransferase NSD1 in human neuroblastoma and glioma. Proc Natl Acad Sci USA 106:21830–21835
143. Kurotaki N, Imaizumi K, Harada N et al (2002) Haploinsufficiency of NSD1 causes Sotos syndrome. Nat Genet 30:365–366
144. Kouzarides T (2007) Chromatin modifications and their function. Cell 128:693–705
145. Smith BC, Denu JM (2009) Chemical mechanisms of histone lysine and arginine modifications. Biochim Biophys Acta 1789:45–57
146. Cole PA (2008) Chemical probes for histone-modifying enzymes. Nat Chem Biol 4:590–597
147. Chuikov S, Kurash JK, Wilson JR et al (2004) Regulation of p53 activity through lysine methylation. Nature 432:353–360

148. Huang J, Perez-Burgos L, Placek BJ et al (2006) Repression of p53 activity by Smyd2-mediated methylation. Nature 444:629–632
149. Subramanian K, Jia D, Kapoor-Vazirani P et al (2008) Regulation of estrogen receptor alpha by the SET7 lysine methyltransferase. Mol Cell 30:336–347
150. Lee JS, Kim Y, Kim IS et al (2010) Negative regulation of hypoxic responses via induced Reptin methylation. Mol Cell 39:71–85
151. Binda O, Boyce M, Rush JS et al (2011) A chemical method for labeling lysine methyltransferase substrates. Chembiochem 12:330–334
152. Islam K, Zheng W, Yu H et al (2011) Expanding cofactor repertoire of protein lysine methyltransferase for substrate labeling. ACS Chem Biol 6(7):679–684
153. Peters W, Willnow S, Duisken M et al (2010) Enzymatic site-specific functionalization of protein methyltransferase substrates with alkynes for click labeling. Angew Chem Int Ed Engl 49:5170–5173
154. Krivtsov AV, Armstrong SA (2007) MLL translocations, histone modifications and leukaemia stem-cell development. Nat Rev Cancer 7:823–833
155. Tachibana M, Sugimoto K, Fukushima T et al (2001) Set domain-containing protein, G9a, is a novel lysine-preferring mammalian histone methyltransferase with hyperactivity and specific selectivity to lysines 9 and 27 of histone H3. J Biol Chem 276:25309–25317
156. Rathert P, Dhayalan A, Murakami M et al (2008) Protein lysine methyltransferase G9a acts on non-histone targets. Nat Chem Biol 4:344–346
157. Wozniak RJ, Klimecki WT, Lau SS et al (2007) 5-Aza-2′-deoxycytidine-mediated reductions in G9A histone methyltransferase and histone H3 K9 di-methylation levels are linked to tumor suppressor gene reactivation. Oncogene 26:77–90
158. Wang R, Zheng W, Yu H et al (2011) Labeling substrates of protein arginine methyltransferase with engineered enzymes and matched S-adenosyl-L-methionine analogues. J Am Chem Soc 133:7648–7651
159. Shah K, Liu Y, Deirmengian C et al (1997) Engineering unnatural nucleotide specificity for Rous sarcoma virus tyrosine kinase to uniquely label its direct substrates. Proc Natl Acad Sci USA 94:3565–3570
160. Liu Y, Shah K, Yang F et al (1998) Engineering Src family protein kinases with unnatural nucleotide specificity. Chem Biol 5:91–101
161. Bishop AC, Ubersax JA, Petsch DT et al (2000) A chemical switch for inhibitor-sensitive alleles of any protein kinase. Nature 407:395–401
162. Adam GC, Cravatt BF, Sorensen EJ (2001) Profiling the specific reactivity of the proteome with non-directed activity-based probes. Chem Biol 8:81–95
163. Barglow KT, Cravatt BF (2004) Discovering disease-associated enzymes by proteome reactivity profiling. Chem Biol 11:1523–1531
164. Barglow KT, Cravatt BF (2006) Substrate mimicry in an activity-based probe that targets the nitrilase family of enzymes. Angew Chem Int Ed Engl 45:7408–7411
165. Rhee SG (2006) Cell signaling. H2O2, a necessary evil for cell signaling. Science 312:1882–1883
166. Terada LS (2006) Specificity in reactive oxidant signaling: think globally, act locally. J Cell Biol 174:615–623
167. D'Autreaux B, Toledano MB (2007) ROS as signalling molecules: mechanisms that generate specificity in ROS homeostasis. Nat Rev Mol Cell Biol 8:813–824
168. Poole LB, Nelson KJ (2008) Discovering mechanisms of signaling-mediated cysteine oxidation. Curr Opin Chem Biol 12:18–24
169. Reddie KG, Carroll KS (2008) Expanding the functional diversity of proteins through cysteine oxidation. Curr Opin Chem Biol 12:746–754
170. Benitez LV, Allison WS (1974) The inactivation of the acyl phosphatase activity catalyzed by the sulfenic acid form of glyceraldehyde 3-phosphate dehydrogenase by dimedone and olefins. J Biol Chem 249:6234–6243
171. Leonard SE, Garcia FJ, Goodsell DS et al (2011) Redox-based probes for protein tyrosine phosphatases. Angew Chem Int Ed Engl 50:4423–4427

172. Leonard SE, Reddie KG, Carroll KS (2009) Mining the thiol proteome for sulfenic acid modifications reveals new targets for oxidation in cells. ACS Chem Biol 4:783–799
173. Jessani N, Young JA, Diaz SL et al (2005) Class assignment of sequence-unrelated members of enzyme superfamilies by activity-based protein profiling. Angew Chem Int Ed Engl 44:2400–2403
174. Hayes BK, Varki A (1989) O-acetylation and de-O-acetylation of sialic acids. Sialic acid esterases of diverse evolutionary origins have serine active sites and essential arginine residues. J Biol Chem 264:19443–19448
175. Li YM, Xu M, Lai MT et al (2000) Photoactivated gamma-secretase inhibitors directed to the active site covalently label presenilin 1. Nature 405:689–694
176. Warburg O (1956) On respiratory impairment in cancer cells. Science 124:269–270
177. Christofk HR, Vander Heiden MG, Wu N et al (2008) Pyruvate kinase M2 is a phosphotyrosine-binding protein. Nature 452:181–186
178. Dang L, White DW, Gross S et al (2009) Cancer-associated IDH1 mutations produce 2-hydroxyglutarate. Nature 462:739–744
179. Menendez JA, Lupu R (2007) Fatty acid synthase and the lipogenic phenotype in cancer pathogenesis. Nat Rev Cancer 7:763–777
180. Mohamed MM, Sloane BF (2006) Cysteine cathepsins: multifunctional enzymes in cancer. Nat Rev Cancer 6:764–775
181. Ramos-DeSimone N, Hahn-Dantona E, Sipley J et al (1999) Activation of matrix metalloproteinase-9 (MMP-9) via a converging plasmin/stromelysin-1 cascade enhances tumor cell invasion. J Biol Chem 274:13066–13076
182. Jessani N, Niessen S, Mueller BM et al (2005) Breast cancer cell lines grown *in vivo*: what goes in isn't always the same as what comes out. Cell Cycle 4:253–255
183. Joyce JA, Baruch A, Chehade K et al (2004) Cathepsin cysteine proteases are effectors of invasive growth and angiogenesis during multistage tumorigenesis. Cancer Cell 5:443–453
184. Taubes G (2008) The bacteria fight back. Science 321:356–361
185. Heal WP, Tate EW (2012) On the Applications of Activity-Based Protein Profiling to Microbial Pathogenesis. Topics in Current Chemistry. Springer, Berlin, Heidelberg. doi: 10.1007/128_2011_299
186. Blum G, von Degenfeld G, Merchant MJ et al (2007) Noninvasive optical imaging of cysteine protease activity using fluorescently quenched activity-based probes. Nat Chem Biol 3:668–677
187. Jedeszko C, Sloane BF (2004) Cysteine cathepsins in human cancer. Biol Chem 385:1017–1027
188. Harbeck N, Alt U, Berger U et al (2001) Prognostic impact of proteolytic factors (urokinase-type plasminogen activator, plasminogen activator inhibitor 1, and cathepsins B, D, and L) in primary breast cancer reflects effects of adjuvant systemic therapy. Clin Cancer Res 7:2757–2764
189. Foekens JA, Kos J, Peters HA et al (1998) Prognostic significance of cathepsins B and L in primary human breast cancer. J Clin Oncol 16:1013–1021
190. Rayo J, Amara N, Krief P et al (2011) Live cell labeling of native intracellular bacterial receptors using aniline-catalyzed oxime ligation. J Am Chem Soc 133(19):7469–7475
191. Krysiak J, Breinbauer R (2012) Activity-based protein profiling for natural product target discovery. Topics in Current Chemistry. Springer, Berlin, Heidelberg. doi: 10.1007/128_2011_289
192. Bulaj G, Kortemme T, Goldenberg DP (1998) Ionization-reactivity relationships for cysteine thiols in polypeptides. Biochemistry 37:8965–8972
193. Lewis CT, Seyer JM, Carlson GM (1989) Cysteine 288: an essential hyperreactive thiol of cytosolic phosphoenolpyruvate carboxykinase (GTP). J Biol Chem 264:27–33
194. Knowles JR (1976) The intrinsic pKa-values of functional groups in enzymes: improper deductions from the pH-dependence of steady-state parameters. CRC Crit Rev Biochem 4:165–173
195. Weerapana E, Wang C, Simon GM et al (2010) Quantitative reactivity profiling predicts functional cysteines in proteomes. Nature 468:790–795

Top Curr Chem (2012) 324: 43–84
DOI: 10.1007/128_2011_289
© Springer-Verlag Berlin Heidelberg 2011
Published online: 25 October 2011

Activity-Based Protein Profiling for Natural Product Target Discovery

Joanna Krysiak and Rolf Breinbauer

Abstract Natural products represent an important treasure box of biologically active molecules, from which many drug candidates have been sourced. The identification of the target proteins addressed by these natural products is a foremost goal for which new techniques are required. Activity-based protein profiling (ABPP), exploiting protein-reactive functional groups present in many natural products, offers unseen opportunities in this respect. This review article describes the current status of this field. Many examples are given for the annotation of biological target proteins of natural products containing epoxides, lactones, lactams, Michael acceptors, and other electrophilic groups. In addition, the development of probe molecules identified from biomimetic natural product libraries is discussed.

Keywords Activity-based protein profiling · Chemical biology · Molecular probes · Natural products · Target profiling

Contents

J. Krysiak and R. Breinbauer (✉)
Institute of Organic Chemistry, Graz University of Technology, Stremayrgasse 9, Graz 8010, Austria
e-mail: breinbauer@tugraz.at

Abbreviations

ABPP	Activity-based protein profiling
ADPKD	Autosomal dominant polycystic kidney disease
AGT	Alkylguanine DNA alkyltransferase
Asn	Asparagine
ATP	Adenosine triphosphate
BG	Benzylguanine
BIOS	Biology-oriented synthesis
BODIPY	Boron-dipyrromethane fluorescent dye
Cys	Cysteine
DNA	Deoxyribonucleic acid
DOS	Diversity-oriented synthesis
FAS	Fatty acid synthase
FDA	Food and Drug Administration
GT	Glutamyl transpeptidase
HSP	Heat shock protein
HUVEC	Human umbilical venous endothelial cells
IC_{50}	Half maximal inhibitory concentration
Lys	Lysine
MALDI	Maxtrix-assisted laser desorption ionization
MIC	Minimal inhibitory concentration
MRSA	Methicillin-resistant *Staphylococcus aureus*
MS	Mass spectrometry
NES	Nuclear export sequence
PAGE	Polyacrylamide gel electrophoresis
PBP	Penicillin binding protein
PIKK	Phosphoinositoide 3-kinase related kinase
PP	Protein phosphatase
RNA	Ribonucleic acid
S	Svedberg unit
SDS	Sodium dodecylsulfate
Ser	Serine
TAMRA	Tetramethyl-5(6)-carboxyrhodamine dye
THL	Tetrahydrolipstatin
Thr	Threonine
UDP	Uridine diphosphate
UMP	Uridine monophosphate

1 Introduction

Natural products are a distinguished class of organic molecules, which have turned out to be an incredible source of biologically active molecules. In contrast to synthetic compounds, they are privileged by their biosynthetic origin and their evolutionary purpose to address (in most cases) protein targets to exert their biological function. The evolutionary relationship between proteins of different species and even different functions might contribute to the explanation of why a natural product such as discodermolide, which is produced by a marine sponge, is also active against breast cancer in humans [1]. In an important analysis, Newman estimated that 47% of the drugs that received clinical approval over the last 25 years are either natural products, derivatives of natural products, or are based on a natural product lead [2].

The current innovation deficit in the pharmaceutical industry has initiated various programs to meet the challenges imposed by new and more difficult drug targets, more complex biology, and the tougher requirements for registration. Although in the recent literature some authors have claimed that for certain drug targets natural products do not perform as well as compound libraries arising from DOS (diversity oriented synthesis) [3], the general view is that natural products have a privileged status due their biosynthetic origin and remain an important source of future drug candidates [4]. As a consequence, two lines of research are pursued in this respect. The first involves inclusion of isolated natural products or natural product extracts in the compound collections used in high-throughput screening campaigns of pharmaceutical companies or academic consortia. The second involves the systematic scientific investigation of traditional folk medicine (traditional Chinese medicine, Ayurveda, ethnomedicine, etc.). The application of modern analytical methods to isolate the active principles of herbal medicines will identify new natural products as drug candidates.

2 Techniques for Identification of Protein Targets

These opportunities for drug discovery are complemented by the significant challenges imposed by the interest in and need for identification of the drug target(s) of these natural products. As the established paradigm of "specific compounds which bind to a single drug target" is now challenged by the insight that many drugs are effective because they modulate the function of several proteins at the same time ("polypharmacology") [5], the problem of target annotation has become relevant for any small molecule drug, independent of its origin (synthetic or natural product) [6–8]. For the identification of drug targets of natural products, the approaches outlined in Fig. 1 are pursued.

Qualified Guess

⊖ is not systematic
⊖ requires a significant amount of available literature information
⊖ will not lead to a comprehensive annotation
⊕ can lead to a result very fast

Identification in Specific Protein Assays

⊖ will not lead to comprehensive annotation
⊕ simple to use; can be performed in a high-throughput format

Identification via Matching Phenocopies

⊖ will not lead to comprehensive annotation
⊖ immature, but emergent technique
⊖ probably not feasible for compounds adressing multiple targets
⊕ is designed for *in vivo* systems

Affinity Based Approach (Chemical Proteomics)

⊖ requires modification of the molecule to introduce an affinity tag
⊖ location of affinity modification might affect biological activity
⊖ cannot be used in *in vivo* systems, but only with cell lysates
⊖ limited to ligands with strong affinity to protein targets ($K_d < 1\mu M$)
⊕ will give a rather comprehensive annotation
⊕ in principle usable for any natural product which binds reversibly to a protein

Activity Based Protein Profiling

⊖ limited to natural products which bind irreversibly to a protein target
⊖ requires modification of the molecule to introduce a labeling tag
⊖ location of affinity modification might affect biological activity
⊕ will give a rather comprehensive annotation
⊕ can be used *in vivo*

Fig. 1 Overview of the different approaches used for target identification of natural products

2.1 Qualified Guesses

Several natural products have been identified as the active principle of herbs or other medicines that have been used against certain diseases in traditional folk medicine. Starting from established mechanistic hypotheses about a certain disease, a natural product active against an illness can be screened for activity against proteins involved in the particular disease mechanism.

2.2 Identification in Specific Protein Assays

High-throughput screening techniques allow the unbiased screening of compound collections (up to three million compounds) in protein or cell-based assays. Routinely, natural products are also part of such screening libraries. Biological data gained from such screens on the one hand allow identification of new tool compounds for new drug targets and, on the other hand, contribute to the annotation of the biological activity of the screened compounds. For example, the public database PUBCHEM (http://pubchem.ncbi.nlm.nih.gov/) makes the screening data of hundreds of performed bioassays available, and increasingly the picture of the protein target profile of chemical compounds and natural products is emerging with higher resolution.

2.3 Identification via Matching Phenocopies

The protein target of compounds can be assigned by comparing the observed phenotype induced by this compound with known phenotypes from comprehensive phenotype collections achieved via gene knockout, RNA interference or by other chemicals [9]. This technique is currently far from maturity, but with increasing progress in automated microscopy, image analysis, etc. it promises a lot for the future. Nevertheless, one should keep in mind that the complexity of a compound that addresses several targets and gives unclear phenotypes will be difficult to resolve.

2.4 Affinity Chromatography of Immobilized Natural Products

This approach involves the immobilization of a natural product (at a site that is not responsible for its biological activity) on an insoluble support (or through a tag such as biotin, which can be used for subsequent immobilization) to form a probe. A cell lysate is brought into contact with this probe, so that any protein that binds to the natural product sticks to the support. After intensive washing of the affinity matrix,

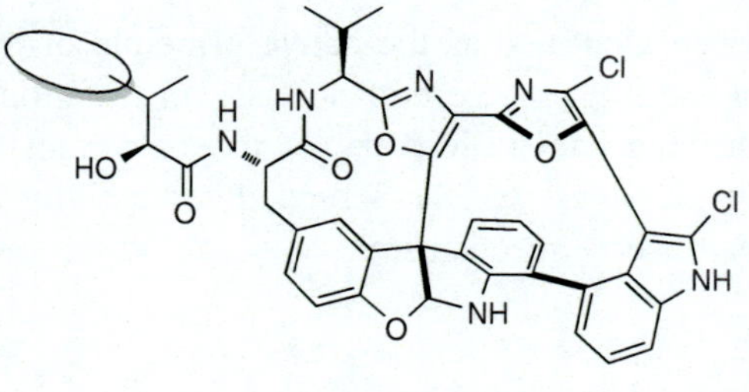

Diazonamide A
1
(δ-Amino transferase)

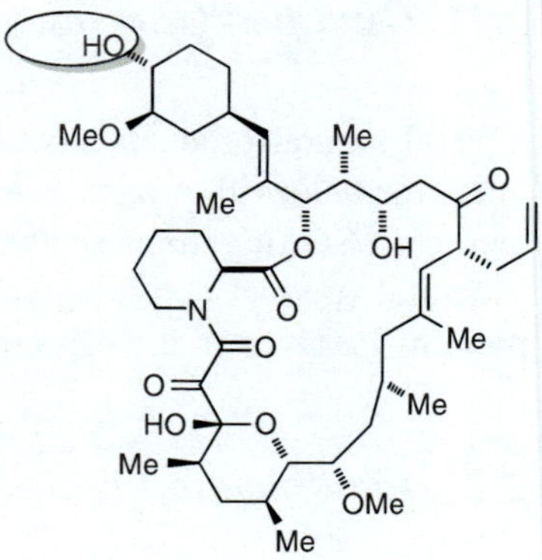

FK 506
2
(FKBP12)

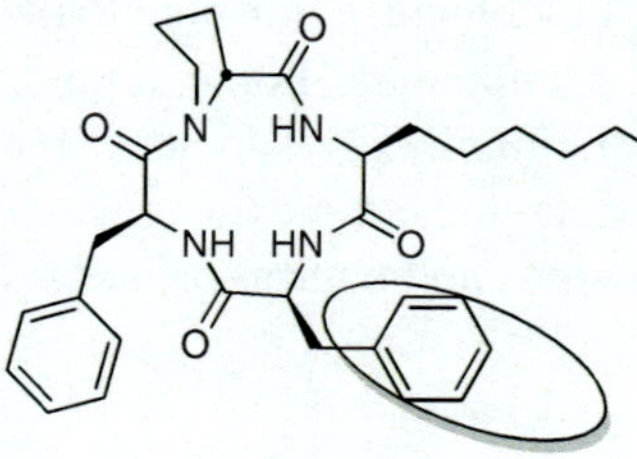

Trapoxin B
3
(Histone deacetylase)

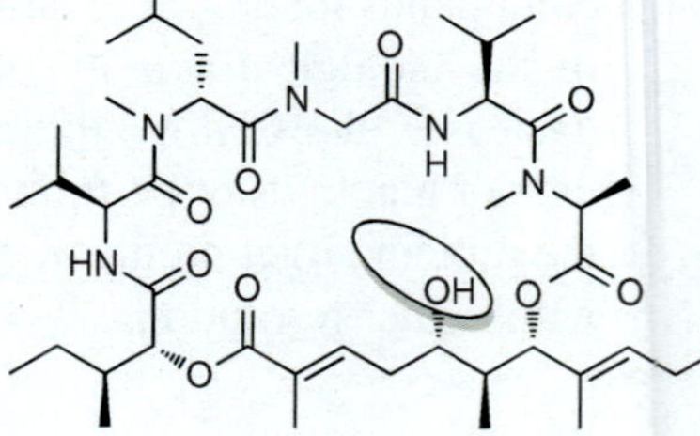

Aurilide
4
(Prohibitin 1 (PHB1))

Fig. 2 Examples of natural products whose targets have been identified by an affinity-based approach. The *ellipse* indicates the location used for immobilization on an affinity matrix

the bound proteins are released, e.g., by addition of a chaotropic reagent. The proteins are then analyzed using standard techniques such as sodium dodecylsulfate polyacrylamide gel electrophoresis (SDS-PAGE) and mass spectroscopy (MS). Control experiments are necessary to distinguish between unspecific and specific binding to the affinity matrix [10].

This technique has led to the target identification of many natural products, such as Diazonamide A (**1**) [11], FK 506 (**2**) [12], Trapoxin (**3**) [13], Aurilide (**4**) [14] and many others (Fig. 2). Taking advantage of the impressive progress in MS has made this approach very popular for annotating the protein targets of drug molecules [15, 16]. This approach requires quite strong binders (typically a dissociation constant, $K_d < 1$ μM is necessary) to produce reliable data, but a modification in which a photoaffinity label is introduced at the probe molecules to enable a covalent linkage has expanded the range of molecules for which this approach is accessible.

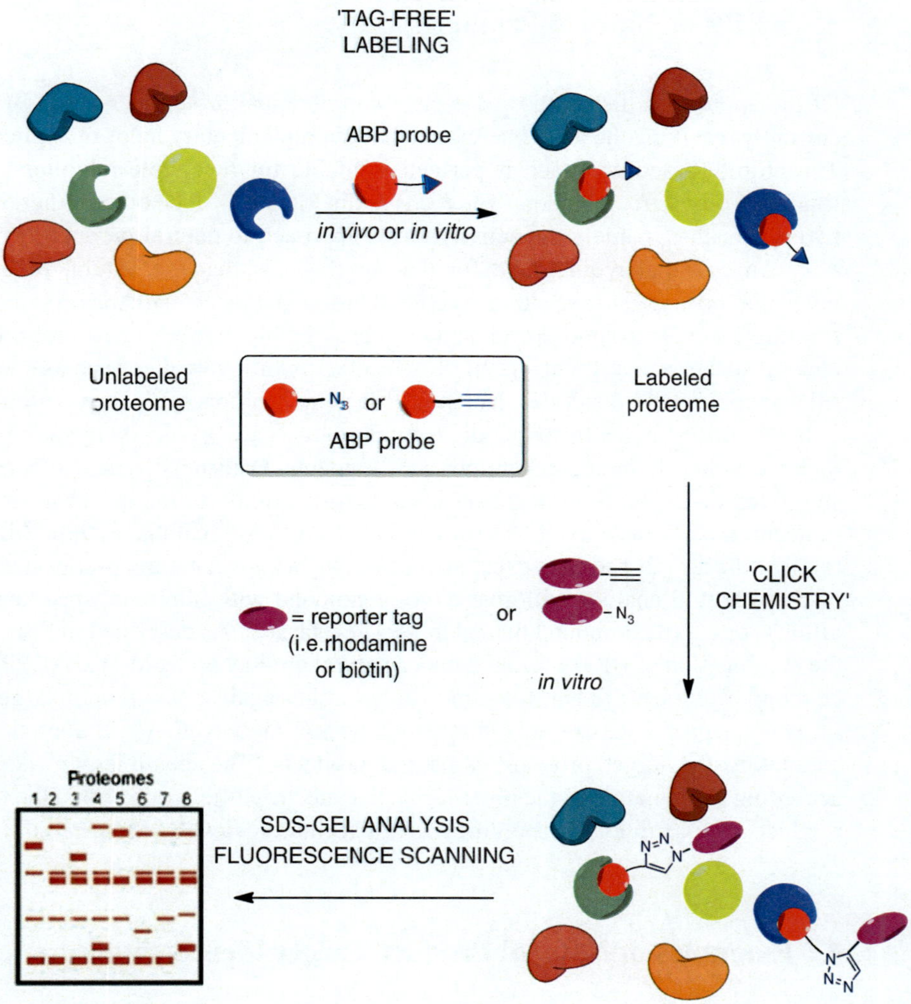

Fig. 3 General workflow in ABPP

2.5 Activity-Based Protein Profiling

Activity-based protein profiling (ABPP) is a technique that builds on a covalent linkage formed between the probe molecule and its protein target (Fig. 3). This technique has been pioneered by Bogyo, Cravatt and others, and has been summarized in several excellent reviews [17, 18]. Typically, the probe molecule contains an electrophilic group, which reacts with a nucleophilic amino acid close to the binding pocket of this substrate. To allow the identification within the complex proteome samples, the probe molecule should contain a label (radioactivity, affinity tag, fluorescence dye) or a group (azide, alkyne, etc.) to which such a label can be introduced via a bioorthogonal ligation reaction.

3 ABPP of Natural Products

Of the approaches listed above the last two (chemical proteomics and ABPP) are currently receiving the most interest and have complementary fields of applications. The affinity-based approach is perfectly suitable for reversible inhibitors, but is limited to quite strong binders. The focus of this chapter will be on the other of these two approaches, namely the activity-based approach to natural product identification. An essential requirement for this approach is that the natural product of investigation contains a reactive functional group that reacts with the protein target, forming a covalent bond. Fortunately, a considerable number of natural products contain such reactive groups [19]. Mostly, electrophilic moieties such as epoxides, Michael acceptors, disulfides, lactones, β-lactams, quinones, etc. can be found.

As described in the Introduction, natural products are privileged by their biosynthetic origin and biological purpose. Waldmann, Danishefsky, and others have suggested that natural products are ideal starting points for the synthesis of probe molecules. The concepts of "biology-oriented synthesis (BIOS)" [20] or "diverted total synthesis" [21] advertise that not necessarily a known natural product but some of its structural analogs will give a tool compound with improved specificity and affinity, or a tool compound that addresses new targets. As described in Sect. 4.3.3, the combination of BIOS around reduced natural product scaffolds and ABPP offers new opportunities for identifying molecular probes against new protein targets.

This chapter discusses several case studies of successful ABPP approaches in annotating the target proteins of natural products. The examples are classified according to the nature of the protein-reactive electrophilic group within the natural products. This article builds on the recent excellent review by Sieber et al. [22].

4 Examples of Natural Product Target Identification by ABPP

4.1 Natural Products Containing an Epoxide Group

4.1.1 E-64

E-64 (L-*trans*-epoxysuccinyl-leucylamido(4-guanidino)butane) **5** was one of the first natural products used as a core for the development of an efficient activity-based probe (Fig. 4). It was isolated from a fungus *Aspergillus japonicus* [23] and its structure comprises an electrophilic epoxysuccinyl motif involved essentially in the inhibition by an active site Cys residue, whereas a Leu unit, which mimics the P2 amino acid of a substrate, is responsible for target recognition, and an agmatine moiety resides in the S3 binding pocket [24]. **5** was found to be a specific irreversible inhibitor of the broad papain family of cysteine proteases [25]. Some of these proteases are involved in physiological processes such as antigen presentation [26], bone remodeling [27], and prohormone processing [28]. Also, several of these

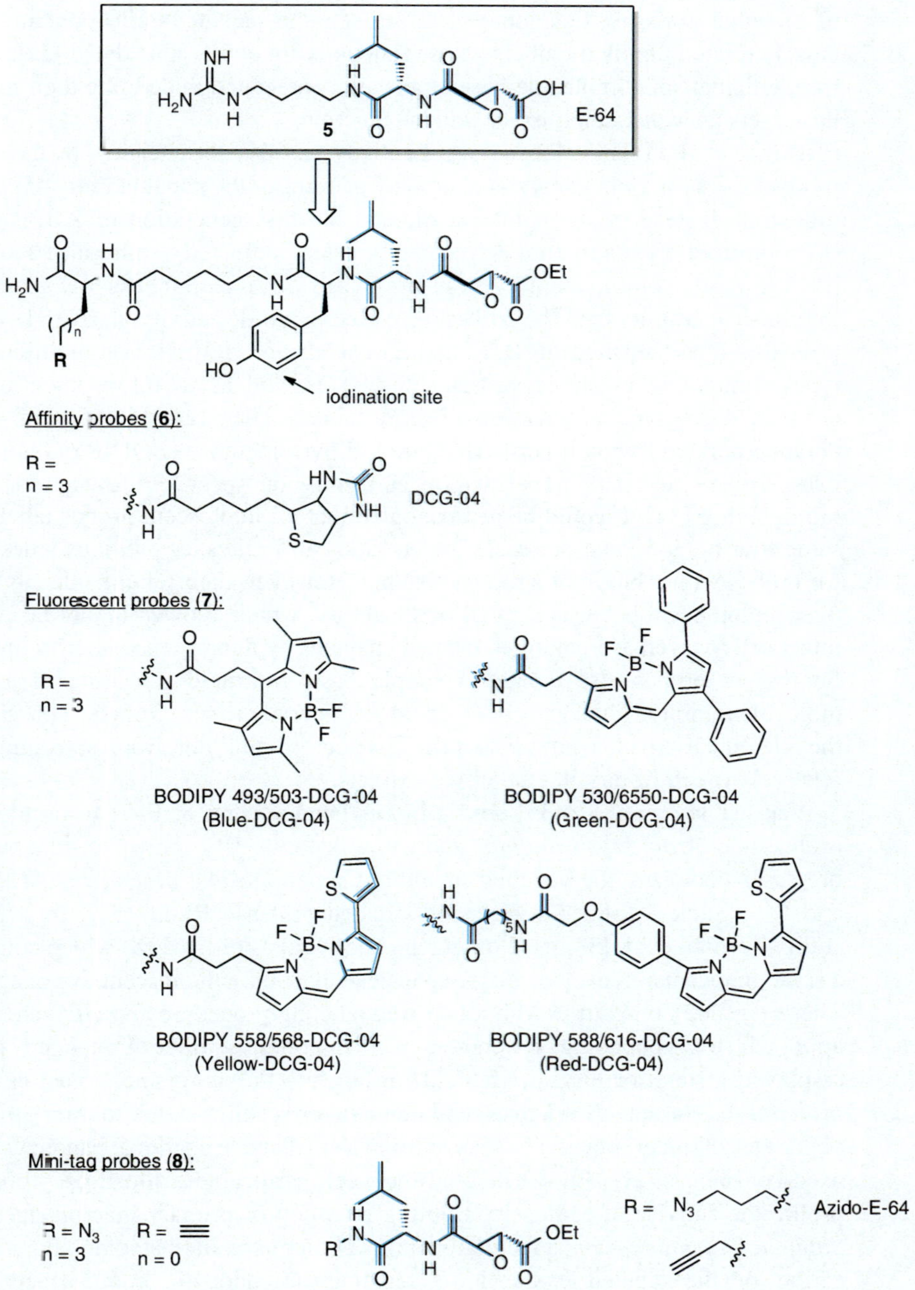

Fig. 4 E-64 and probe derivatives used in ABPP

enzymes have been identified as playing an important role in many pathological conditions, such as osteoporosis [27], asthma [29], rheumatoid arthritis [30], or cancer invasion and metastasis [31, 32], making them a potentially important class of enzymes for drug development. However, the papain family contains many closely related family members whose functions are poorly elucidated [27]. Therefore, efficient small molecule-based tools are appreciated in defining their molecular targets as well as their exact cellular function.

Bogyo et al. [33] developed a set of activity-based probes inspired by the natural product E-64 for global analysis of activity patterns of the papain family of cysteine proteases. Instead of the agmatine moiety, the first generation of ABPP probes (**6**) comprised a Tyr structure, which can be easily iodinated to introduce a radioactive tag. Also, a biotinylated Lys was introduced via an amino hexanoic acid spacer to furnish an affinity tag. The probes were used to profile activity of several cysteine proteases such as cathepsins B, H, and L in dendritic cell lysate and multiple tissue types (brain, kidney, liver, prostate, testes) as well as to follow their activity changes during the progression of skin cancer. The second library of probes (**7**) incorporated four cell-permeable boron-dipyrromethane (BODIPY) fluorescent dyes with nonoverlapping excitation and emission spectra, allowing for probe multiplexing [34]. It could be demonstrated that all fluorescent probes labeled the same four predominant proteases in dendritic cell lysates, as previously described for probes with a biotin or a radioactive tag. However, their remarkable advantage over the other probes was the cell permeability, which allowed in situ labeling in intact cells as well as imaging of labeled enzymes by fluorescence microscopy. The fluorescent probes also created a simple basis for rapid screening of a small molecule inhibitor library for selected targets in crude tissue extracts. This allowed the identification of a new cathepsin B-selective inhibitor from screening of a relatively broad library in crude liver extracts.

Van der Hoorn et al. [35] used a biotinylated version of E-64 to profile plant proteases in *Arabidopsis thaliana* and was able to identify six papain-like cysteine proteases enclosing the vacuolar *Arabidopsis* aleurain-like protease (AALP) [36] and the vesicle-localized desiccation-induced protease RD21 [37]. In a further study, Kashani et al. [38] used minitagged derivatives of E-64 (**8**) which contained a small innocent alkyne or azide group instead of a bulky fluorescent dye or a biotin. These tags are compatible with a two-step labeling procedure and click chemistry approach for appending a reporter tag. Importantly, the minitagged probes displayed a superior potency (tenfold) in labeling the above-mentioned enzymes in vivo (i.e., on detached leaves, cell cultures, or seedlings) due to their ability to penetrate cell membranes. The idea of using a small azide chemical handle was also used by Hang et al. [39]. They used their cell-permeable azido-E-64 probe **8** to profile the activity of cathepsin B during infection of primary macrophages with *Salmonella typhimurium*, a facultative intracellular bacterial pathogen. For visualization of the labeled cathepsin B residing in endocytic compartments, the Staudinger ligation was employed to attach a biotin tag. Interestingly, it could be demonstrated that the active cathepsin B is specifically excluded from *Salmonella*-containing vacuoles in infected cells, leading to the conclusion that the inhibition of endocytic proteases may give rise to the survival and virulence of intracellular bacterial pathogens.

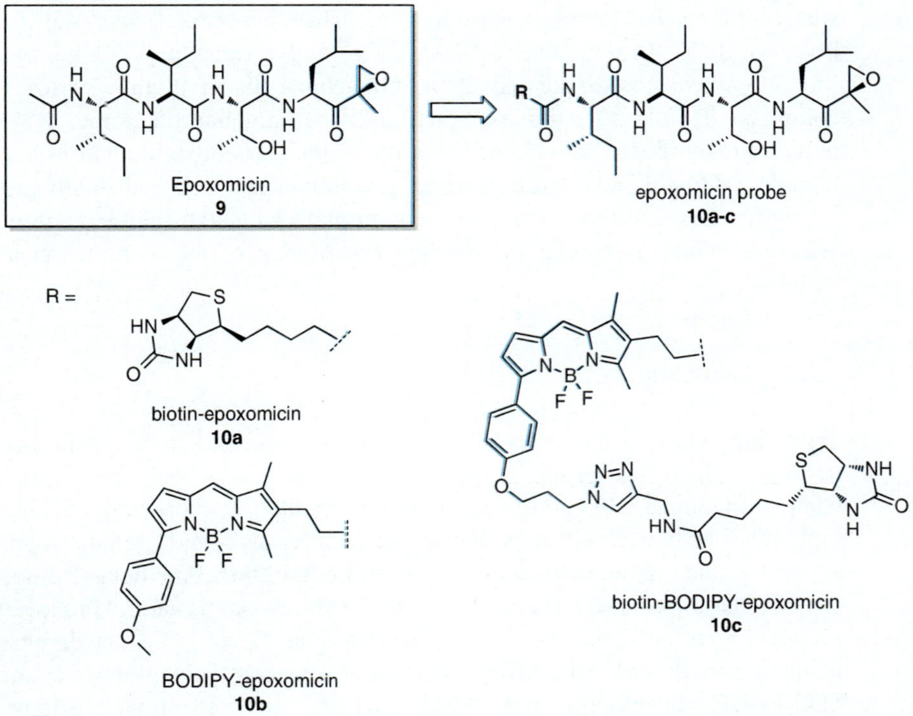

Fig. 5 Epoxomicin and probe derivatives used in ABPP

4.1.2 Epoxomicin

Epoxomicin (**9**) is a natural product comprising an epoxy-β-aminoketone moiety and tripeptide side chain and was isolated from an *Actinomycetes* strain [40]. It exhibits a strong antitumor activity against B16 mouse melanoma in vivo. To clarify the mode of action leading to this biological effect, the research group of Crews prepared a synthetic biotinylated derivative **10a** of epoxomicin (Fig. 5) to reveal its molecular targets [41, 42]. Incubation of the epoxomicin probe with the murine lymphoma EL4 cell line led to the detection of four catalytically active β subunits of the 20S proteasome: LMP7, LMPX, LMPZ, and MECL1. Thus, epoxomicin inhibits primarily the chymotrypsin-like activity of the proteolytic machinery of the 20S proteasome. It was also found that epoxomicin efficiently inhibited the nuclear factor-κB (NF-κB) regulator of inflammation and was able to reduce significantly a cutaneous inflammation in a mouse model. In further ABPP studies, BODIPY-derived epoxomicin probes **10b** and **10c** confirmed that epoxomicin inhibits effectively all three β subunits of the constitutive proteasome (β1, β2, β5) as well as of the immunoproteasome (β1i, β2i, β5i) in mammalian cells [43] and in plants (*A. thaliana*) [44]. Importantly and interestingly, it was demonstrated that epoxomicin, in contrast to other proteasome inhibitors (i.e., vinyl sulfones, lactacystin, bortezomib, NC005), was able to block the activity of the β5t subunit of murine thymoproteasome [45]. The thymoproteasome was found to be a third 20S proteasome particle, next to the constitutive proteasome and immunoproteasome, expressed in the epithelial cells of thymus cortex [46]. It comprises β1i and β2i subunits, exactly like the immunoproteasome, as well as a unique subunit β5t of unexplained activity [47]. Using epoxomicin probes, it could be revealed that β5t exhibits an intrinsic proteasome activity, exclusively present in young thymus, and that this unique subunit plays an active role in positive T cell selection, which is crucial for efficient functioning of the mammalian immune system.

4.1.3 Triptolide

Triptolide (**11**) is a diterpene triepoxide (Fig. 6) isolated from the *Tripterygium wilfordii*, which is a medicinal plant whose extracts have been used in traditional Chinese medicine as *lei gong teng*. Triptolide exhibits a whole range of interesting biological activities, such as potent antiproliferative and immunosuppressive activities. The proven medicinal track record in traditional medicine for important diseases has raised questions about the protein targets of this natural product. Crews prepared a tritium-labeled version ([^{3}H]triptolide, **12**) (Fig. 6) that they used for incubation studies of cells. After "extensive chromatographic protein fractionation, SDS-PAGE separation, MALDI-MS analysis" they identified, among other proteins of the membrane fraction, PC2 (calcium channel polycystin-2), whose cellular function they studied further. They could demonstrate that triptolide induces Ca^{2+} release by a PC2-dependent mechanism. Interestingly, they could show in a murine model of autosomal dominant polycystic kidney disease (ADPKD), an inherited disease that ultimately leads to renal failure, that triptolide attenuates overall cyst growth by restoring Ca^{2+} signaling in these cells [48]. In the same year, a research group at Merck using a similar approach but focusing on the anti-inflammatory effect of triptolide reported that this natural product covalently binds to a 90-kDa nuclear protein [49]. Recently, this protein was identified

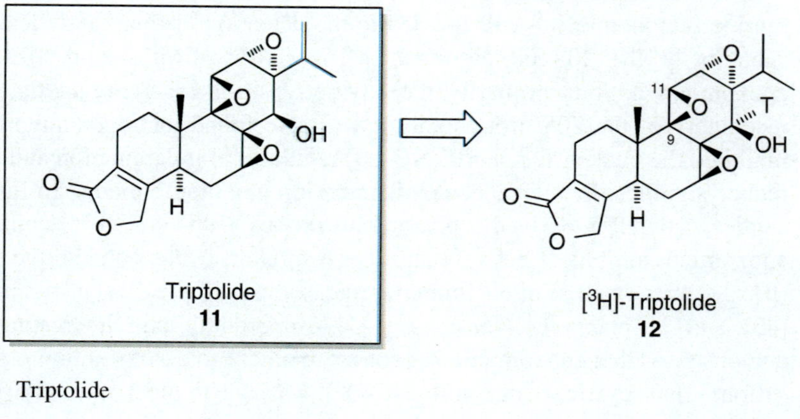

Fig. 6 Triptolide

as XPB, a subunit of the transcription factor TFIIH. Triptolide inhibits its DNA-dependent ATPase activity, which results in inhibition of RNA polymerase II-mediated transcription and nucleotide excision repair [50]. Triptolide features several potentially reactive groups, such as an α,β-unsaturated γ-lactone and three consecutive epoxides. The studies related to XPB indicate that the 9,11-epoxide is responsible for the covalent binding. Further studies will be necessary to give a full description of the interesting biological functions of this molecule.

4.1.4 Fumagillin

The antiangiogenic natural product fumagillin (**13**) inhibits neovascularization via endothelial cell cycle arrest in the late G1-phase. Crews at al. speculated that the epoxide groups in this molecule might result in a covalent interaction with its target protein (Fig. 7) [51]. Having an activity-based approach in mind, they converted natural fumagillin with a few straightforward transformations into the biotinylated probe **14**, which retained cytostatic activity toward human umbilical vein endothelial cells (HUVECs). Incubation of HUVECs with **14**, followed by adsorption to streptavidin agarose, SDS-PAGE and blotting with avidin–horse radish peroxidase revealed a newly biotinylated 67-kDa band. Considerable efforts were necessary to isolate sufficient protein material for Edman sequencing, but this was ultimately rewarded by the identification of human methionyl aminopeptidase (metAP-2) as the cellular target of fumagillin [51].

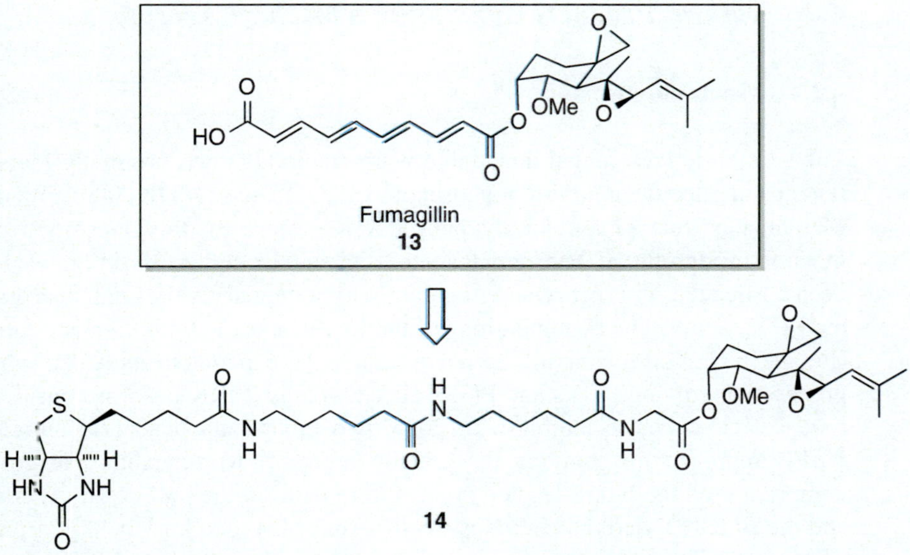

Fig. 7 Fumagillin and biotinylated probe derivative

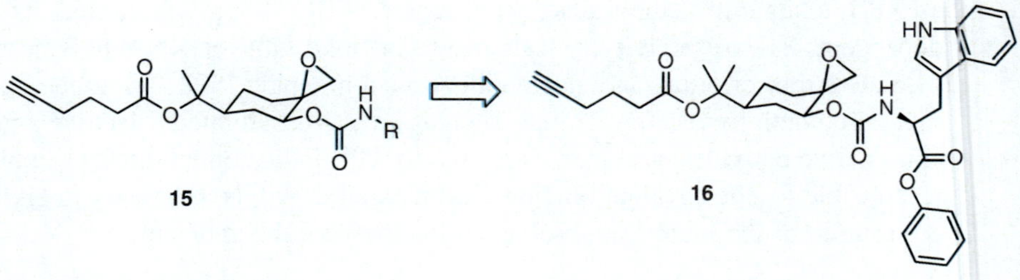

Fig. 8 Spiroepoxide library

4.1.5 Biomimetic Spiroepoxide Library

Inspired by the spiroepoxide structure of natural products such as fumagillin (**13**), lumanicin D, and FR901464, the groups of Cravatt and Sorensen designed a ~50-member library of spiroepoxides (**15**) containing an amino-derived recognition unit and clickable alkyne tag (Fig. 8) [52]. This library was screened for antiproliferation activity against the human breast cancer line MDA-MB-231. Compound **16** showed the highest activity and was also the compound that was used in an ABPP approach to identify its protein targets. Notably, brain-type phosphoglycerate mutase 1 (PGAM1) was identified as a protein target of **16**. This protein is a key enzyme in glycolysis and converts 3-phosphoglycerate to 2-phosphoglycerate. Interestingly, labeling and inhibition of PGAM1 with **16** was observed exclusively in intact cells [52].

4.2 Natural Products Containing a Michael Acceptor

4.2.1 Wortmannin

The steroid-derived fungal metabolite wortmannin (**17**) was originally found as a potent and selective inhibitor of phosphoinositide 3-kinase (PI3K) family members. Members of this kinase family phosphorylate the 3′-hydroxyl position of the inositol headgroup of phosphoinositides lipids. The mode of action of **17** has been associated with irreversible binding with a crucial Lys residue conserved in the PI3K family. The ε-amino group of the Lys attacks the furan ring and forms **18** (Fig. 9) [53]. Wortmannin has been used as a tool compound to study cell signaling processes involving PI3Ks and PI3K-related kinases (PIKK), as it is considered to be a selective kinase inhibitor. In 2005, two groups independently reported an ABPP approach for studying the protein targets of wortmannin. The group of Cimprich used the biotin-labeled probe **19a** to perform the pull-down experiments and the BODIPY derivative **19b** as a cell-permeable probe for labeling of proteins within cells. They could confirm certain members of the PI3K and PIKK families, such as ATR, ATM, DNA-PKc, and a protein entitled p110 as targets of

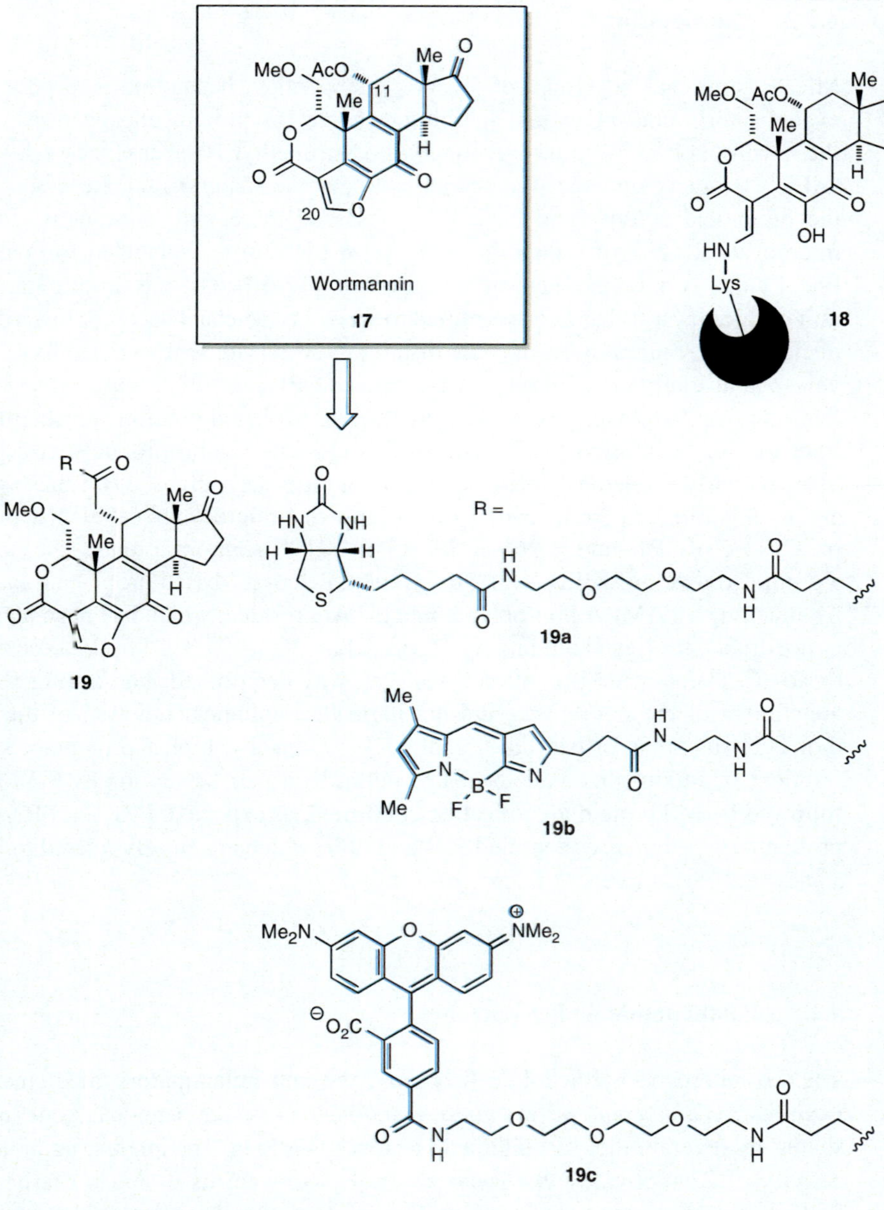

Fig. 9 Wortmannin and probe derivatives used in ABPP

wortmannin [54]. In parallel, the group of Liu used the TAMRA probe **19c** for similar purposes [55]. Interestingly, in addition to the expected PI3Ks, they identified members of the polo-like kinase (Plk1) family as protein targets potently inhibited by wortmannin. This finding is quite important as, on the one hand, it documents the broader range of kinase families addressed by wortmannin and, on the other hand, provides with **19c** a tool compound that allows the monitoring of Plk1 expression during the cell cycle [55].

4.2.2 Microcystins

Microcystins are a group of cyclic heptapeptide hepatotoxins produced by cyanobacteria and are potent inhibitors of Ser/Thr protein phosphatases. While the hydrophobic 3-amino-9-methoxy-2,6,8-trimethyl-10-phenyldeca-4,6-dienoic (ADDA) moiety and the *N*-methyldehydroalanine (Mdha) unit are essential for the biological activity and have been observed to remain constant in different microcystins, the Arg residue in microcystin LR (**20**) is exchanged for Val, Met, Phe, Leu, or Ala in other natural microcystins (Fig. 10). The α,β-unsaturated Mdha unit of microcystins has been identified to serve as a Michael acceptor towards thiol nucleophiles. Surprisingly, adducts of microcystins with aminoethanethiol remain very potent inhibitors of protein phosphatases PP1 and PP2A, suggesting that the hydrophobic ADDA group is a key contributor to ligand binding. To shed further light into the inhibition mechanism of this class of natural products, two groups independently prepared [^{125}I]-labeled microcystin derivatives. With these probes and by mutation experiments they could show that microcystin covalently binds to the Cys273 of PP1 and Cys266 of PP2A [56, 57]. This information set the stage for a group from ActivX Biosciences to synthesize activity-based probe compound **21** by attaching a TAMRA fluorophore onto the Arg residue, which had been identified as not crucial for ligand binding. For probe **21**, an IC_{50} of 4.0 nM for PP1 (compared to an IC_{50} of 0.3 nM for microcystin **20**) was determined, confirming that the attachment of the fluorophore did not harm the biological function of the probe molecule. In order to test the use of probe **21** in a complex proteome, soluble fractions of Jurkat cells were incubated with **21**. Protein separation by SDS-PAGE followed by MS protein identification confirmed, as expected, PP1 and PP2A to be protein targets, but also identified PP4 and PP6 (which are closely related to PP2A) as new covalent targets of microcystin [58].

4.2.3 Parthenolide and γ-Lactones

The sesquiterpene parthenolide (**22**) from the anti-inflammatory medicinal herb Feverfew (*Tanacetum parthenium*) embodies a α-methylene-γ-lactone moiety shared by several other sesquiterpene lactones (Fig. 11). The interesting biological activity of **22** motivated the group of Crews to reveal its biological target [59]. Parthenolide possesses two electrophilic moieties, namely an internal epoxide and an α,β-unsaturated lactone. Reduction of the exocyclic double bond of **22** furnished reduced parthenolide **23**, which showed none of the anti-inflammatory activity characteristic of the parent compound. Having confirmed the importance of the α-methylene-γ-lactone, the Crews group set out to use the biotinylated parthenolide probe **24** for an affinity-based pull-down experiment. Affinity purification of biotinylated proteins using streptavidin resin, followed by SDS-PAGE and western blotting demonstrated that parthenolide formed a covalent adduct with IκB kinase β (IKKβ), a protein that is involved in the cascade activating the NF-κB. Tryptic

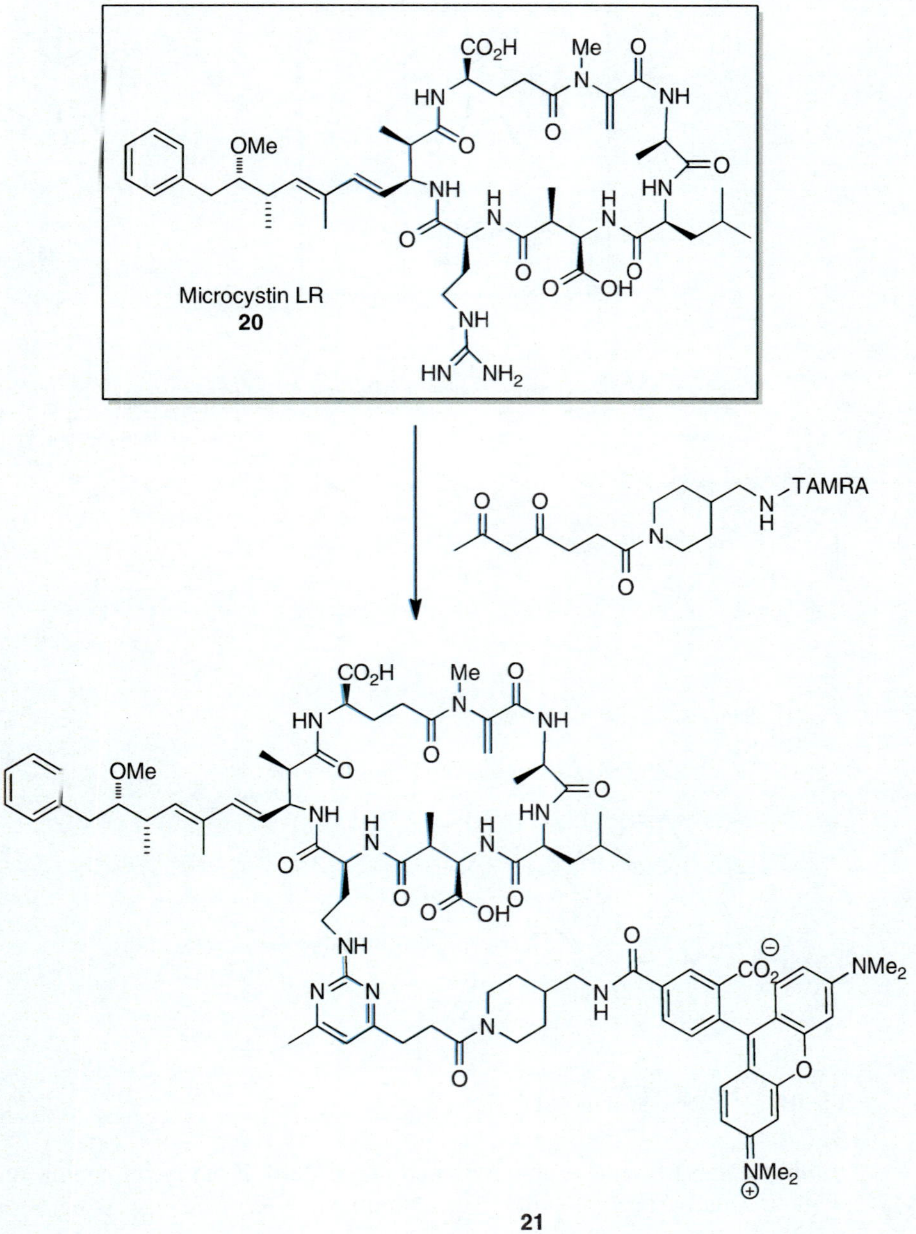

21

Fig. 10 Microcystin LR

digest and MS analysis complemented by mutation studies clearly showed that the biological action of parthenolide results from a Michael-type addition reaction of Cys179 of IKKβ at the exocyclic methylene group of the lactone ring [59].

Very recently, Sieber et al. investigated in an important study the general suitability of γ-lactones (being present in almost 10% of all natural products) and

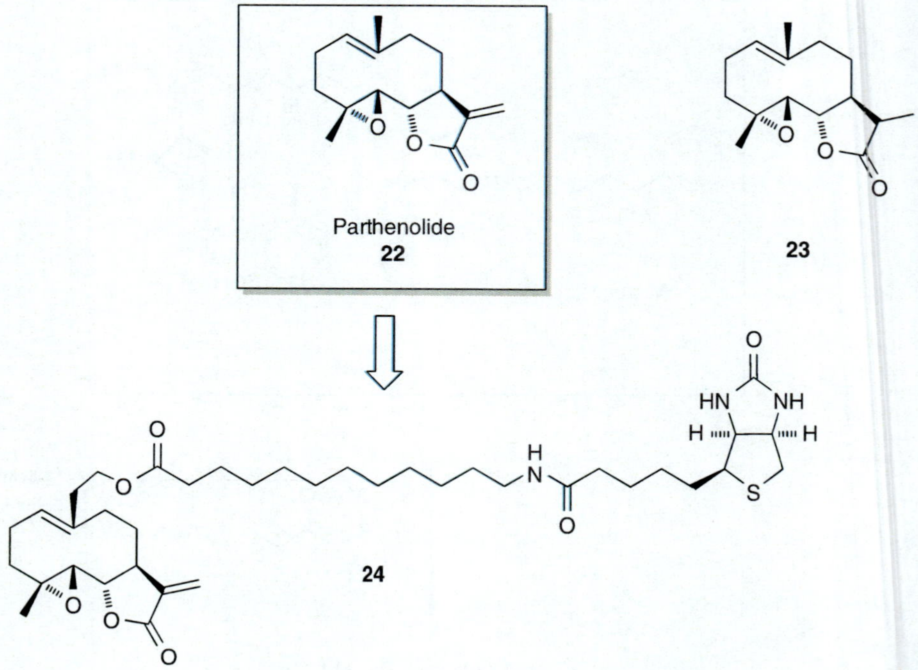

Fig. 11 Parthenolide and biotinylated derivative

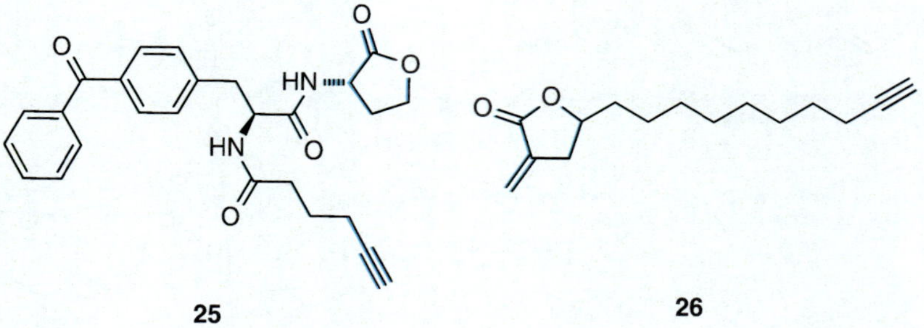

Fig. 12 γ-Lactones used in ABPP

γ-thiolactones to serve as activity-based probes [60]. From the proteomic analysis of their small, but well-chosen set of minimal γ-lactones possessing an alkyne handle for labeling purposes, they could conclude that these molecules showed only little reactivity with proteomic samples. Equipping their probe **25** with a benzophenone unit suitable for photocrosslinking allowed them to identify protein targets that noncovalently interact with lactone molecules (Fig. 12). Incubating bacterial proteomes with **25** and subsequent photocrosslinking at 366 nm presented trigger factor, a protein that has peptidylprolyl isomerase activity, as a target of this molecule. In contrast, the α-methylene-γ-lactone moiety, which is present in 3% of all natural products, serves as an excellent electrophilic trap and makes these molecules suitable for ABPP analysis, as discussed for parthenolide (Sect. 4.2.3). Sieber and coworkers used the minimal scaffold probe **26** to search for covalently addressed targets of α-methylene-γ-lactones in bacterial proteomes. Formate acyltransferase, dihydrolipoamide dehydrogenase, 6-phosphofructokinase, MurA1, and MurA2 (the latter two enzymes being essential for bacterial cell wall biosynthesis) were identified via MS analysis [60].

4.2.4 Phoslactomycin

Phoslactomycins A–F are antifungal antibiotics isolated from *Streptomyces* sp. and reported to be potent protein Ser/Thr phosphatase 2A (PP2A) inhibitors in vitro. The α,β-unsaturated δ-lactone moiety should be sufficiently electrophilic to serve as an anchor point for covalent attachment to the target proteins of phoslactomycin. The group of Simizu tested this hypothesis by attaching to phoslactomycin A (**27**) a biotin label furnishing probe molecule **28** (Fig. 13) [61]. Human fibrosarcoma HT1080 cells were incubated with probe **28** for 30 min, lysed, and the

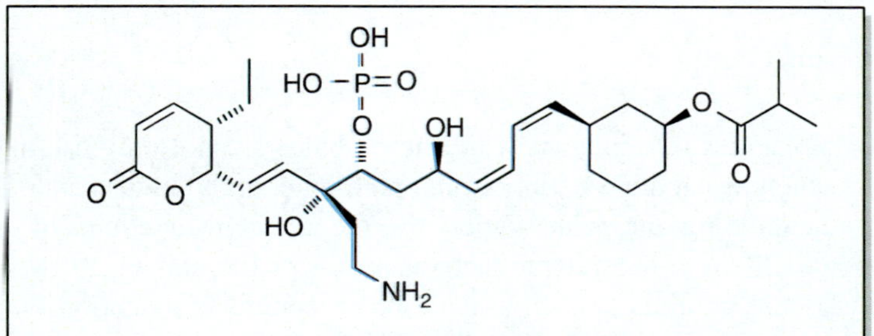

Phoslactomycin A
27

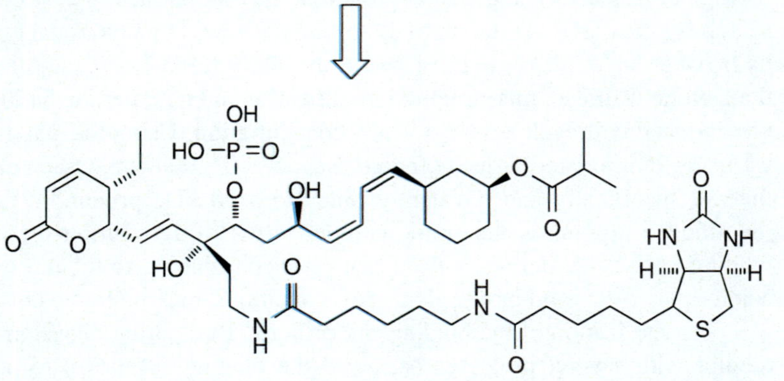

28

Fig. 13 Phoslactomycin and biotinylated derivative

biotin-containing complexes isolated by the use of streptavidin-conjugated agarose beads. The **28**-bound proteins were analyzed by SDS-PAGE followed by silver staining. Compared with appropriate control probes, five additional bands were found in the gel (120 kDa, 65 kDa, and three 36 kDa bands). Western blotting analysis confirmed that the bands at 36 kDa account for the catalytic subunits PP2Ac, PP2Acβ and PP6 of the PP2A family of proteins. Further investigation revealed that the Cys269 of the PP2A catalytic subunit is responsible for the covalent binding to the α,β-unsaturated lactone of phoslactomycin A via a Michael-type addition mechanism. The 65 kDa protein was PP2A and the identity of the 120 kDa protein has not yet been determined but the authors speculate that it might be a protein that associates with one of the PP2A isoforms as a substrate or regulator [61]. Interestingly, the same group later used phoslactomycin as a test case to demonstrate the applicability of a general strategy to immobilize biologically active molecules for pull-down assays using a site-nonselective carbene addition/insertion upon irradiation of diazirine-modified agarose beads [62]. This interesting approach promises to solve one of the greatest bottlenecks of preparing affinity matrixes: providing a fast and easy access to an immobilized version of a natural product without destroying its pharmacological function.

4.2.5 Pironetin

Small molecules that interact with microtubules are not only useful as antitumor agents (such as paclitaxel, vinblastin, epothilone, etc.) but also as tool compounds for understanding the wide variety of the cellular functions of microtubules. Pironetin (**29**) is a natural product that disrupts the cell cycle by inhibiting the M phase and exhibits antitumor activity by apoptosis induction via microtubule disassembly. The characteristic α,β-unsaturated δ-lactone moiety suggested that it could interact with its target protein in a covalent fashion. This proposal was supported by the observation that its saturated analog showed much weaker biological activity. The group of Osada et al. used an activity-based approach to search for the molecular target of **29** [63]. Attachment of a biotin label at the central hydroxy group of pironetin furnished probe **30** (Fig. 14). Despite its close vicinity to the lactone warhead, this structural modification did not alter the biological activity and induced cell cycle arrest at 1 μM concentration. Using the standard workflow for using biotinylated probe compounds as described above, Osada et al. observed that **30** specifically and covalently binds to a 50 kDa protein. Western blotting identified α-tubulin as the single cellular target of **29**. During the course of the binding assay, the authors noticed that pironetin binding to tubulin is labile under acidic ($pH < 4.5$) and basic ($pH > 8.0$) conditions, making liquid chromatography-MS characterization of the binding site difficult. But, partial digestion of **30**-treated tubulin with several proteases narrowed the binding site down to being between residues 270 and 370 of α-tubulin. As the addition of a Cys or Lys nucleophile across the α,β-unsaturated δ-lactone moiety seemed to be a plausible mode of action, an Ala exchange with all Cys and Lys residues within this region was performed. From 12

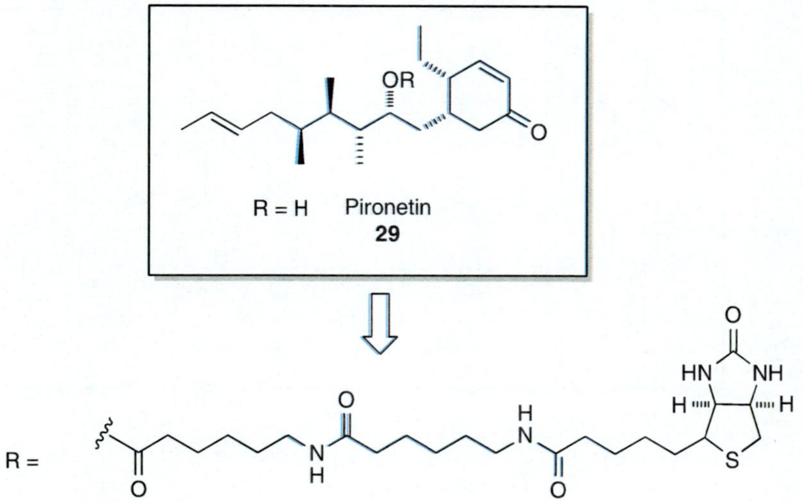

30

Fig. 14 Pironetin and biotinylated derivative

such mutants only Lys352 emerged as the crucial amino acid responsible for the binding of pironetin to α-tubulin, although five Cys residues with intrinsically higher nucleophilicity were also found in this protein region [63]. This finding emphasizes that not only the reactive group but also the remaining structure of the natural product are instrumental in achieving selectivity in the binding of the probe to its protein targets.

4.2.6 Leptomycin B

The transport of cellular proteins between nucleus and cytoplasm across the nuclear envelope is mediated by energy-dependent transport. Specific sequences within a protein contain the information necessary for the nucleocytoplasmic transport. In lower and higher eukaryotes, CRM1/exportin 1 is the receptor for short nuclear export sequences (NES), which are rich in Leu. The polyketide natural product leptomycin B (**31**) was discovered as a potent antifungal antibiotic that blocks the eukaryotic cell cycle (Fig. 15). Kudo et al. discovered that leptomycin B exerts its function by directly binding to CRM1 and therefore blocking the nuclear export of proteins [64]. After preparing the biotinylated probe **32** from **31** they cultured HeLa cells in the presence of 10 nM **32** for 2 h. The biotin-containing complexes were isolated by streptavidin-conjugated agarose beads and identified by SDS-PAGE and western blotting. Only a single protein band at 102 kDa was identified as being selectively bound by **32**. This protein was identified as CRM1. Further analysis revealed that Cys529 covalently binds to leptomycin B via nucleopilic 1,4-addition across the α,β-unsaturated δ-lactone unit of this natural product [65]. The covalent modification of CRM1 by **31** apparently induces a major change in its

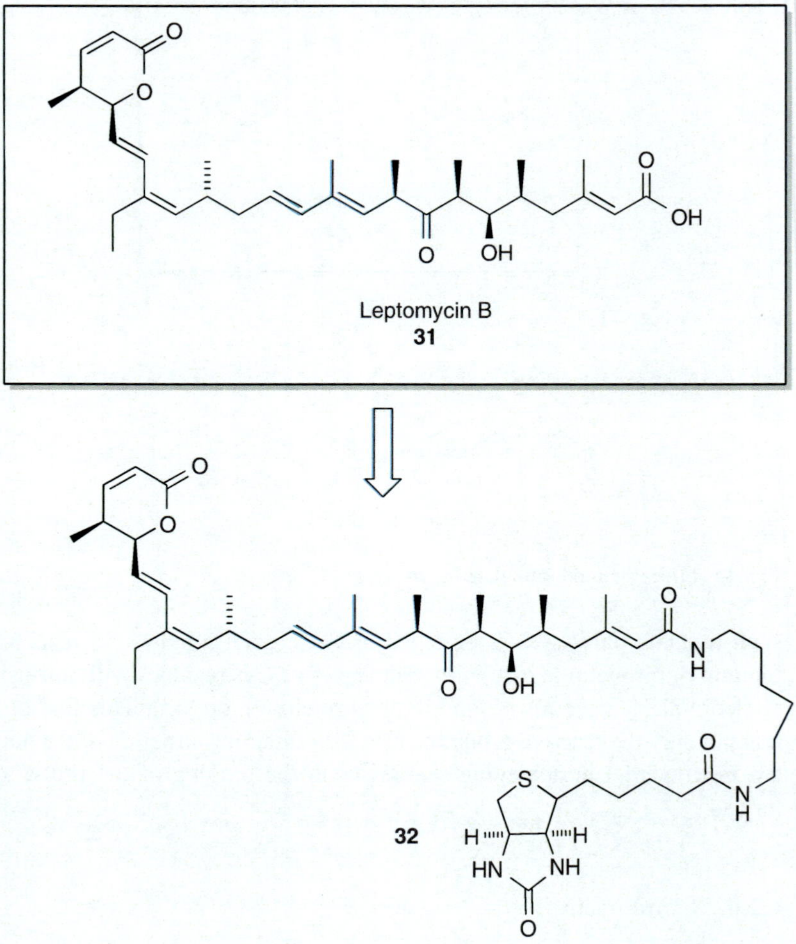

Fig. 15 Leptomycin B and biotinylated derivative

three-dimensional structure in a native gel, and a significant mobility shift is observed. It is fascinating that despite the high in vivo reactivity of **31** with thiol nucleophiles, in the environment of a cell leptomycin B exerts its biological function even at low concentration. The authors speculate that the large hydrophobic part of **31** might prevent access of the more abundant L-cysteine or glutathione to the warhead. Leptomycin has because of its high selectivity seen many applications as a tool compound to study the nuclear export of proteins.

4.2.7 Cyclostreptin [(−)-FR182877]

Because of its complex and unprecedented structure (−)-FR182877 (**33**), which has been renamed cyclostreptin, has attracted the attention of synthetic chemists.

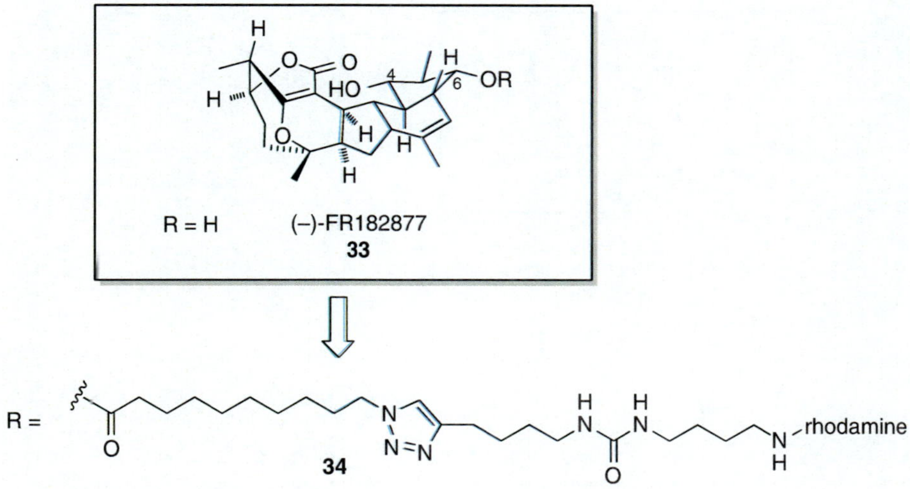

Fig. 16 Cyclostreptin [(−)-FR182877]

Sorensen et al. presented the first total synthesis of this fascinating molecule [66] and later joined forces with the group of Cravatt to search for its protein targets. **33** displays two potential electrophilic sites at which protein nucleophiles could react: the strained bridgehead olefin and the carbonyl group of a lactone (Fig. 16). Sorensen and Cravatt exploited the better accessible 6-OH group to attach a ω-azido alkanoic acid. The azido group allowed the flexible introduction of different labels (rhodamine dye, biotin, etc.) via copper-catalyzed Huisgen–azide alkyne coupling. Rhodamine-labeled probe **34** was used for an ABPP search for the protein target. Proteomic analysis of different mouse tissues identified only one major protein band at 70 kDa as specific for **33**, as compared with other control experiments featuring notably the enantiomer of **33**, which was accessible through the previous total synthesis efforts. Tryptic digestion and MS identification presented carboxylesterase-1 (CE1) as the molecular target of **33** [67]. At this stage, the authors could not explain the very large concentration difference between the inhibition of CE1 ($IC_{50} = 34$ nM) by **33** and the reported effects on tubulin polymerization (~100 μM), and suggested that it would require more detailed investigations. A few years later, Diaz et al. used an MS-based approach to search for protein fragments containing cyclostreptin and demonstrated that cyclostreptin covalently binds to Thr220 and Asn228 of β-tubulin [68]. Their carefully carried out study demonstrated that cyclostreptin favors microtubules (and not unpolymerized tubulin) as the target of the compound, and that the strained olefin is required for cyclostreptin activity [68]. This different outcome using two different activity-based proteomic techniques allows the humble conclusion that neither technique represents a panacea for obtaining comprehensive binding data and that the observed results are also probably influenced by the concentration of compounds and by the state of the investigated cellular systems.

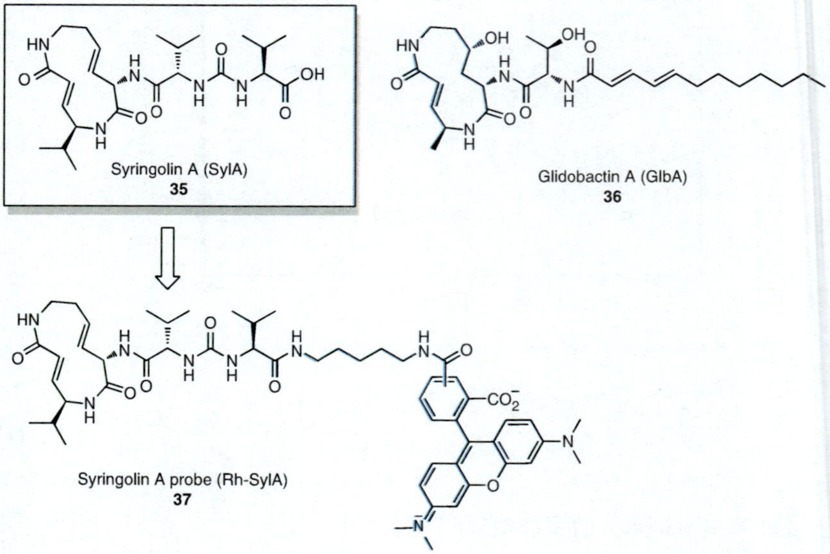

Fig. 17 Syringolin A and glidobactin A

4.2.8 Syringolin A

Syringolin A (SylA, **35**) is a plant effector produced by the phytopathogenic bacterium *Pseudomonas syringae* pv. *Syringae* (*Pss*) [69]. Along with the structurally related natural product glidobactin A (GlbA, **36**) isolated from unknown species of the order *Burkholderiales*, it belongs to the family of the syrbactins. SylA is an unusual peptide product of mixed nonribosomal peptide/polyketide synthetase that contains a 12-membered macrocycle consisting of two nonproteinogenic amino acids 5-methyl-4-amino-2-hexenoic acid and 3,4-dehydrolysine, as well as a dipeptide tail linked through an ureido bond (Fig. 17). To shed light on the role of SylA in the plant–pathogen interaction, a SylA-deletion mutant was generated by targeted gene disruption in a *Pss* strain, which causes brown spot disease on bean. Infection experiments on bean plants showed a significant reduction of virulence in mutant strain, thus identifying SylA as a potent virulence factor [70]. Strikingly, in the same study, the research group of Groll demonstrated that SylA leads to the efficient and irreversible inhibition of eukaryotic 20S proteasome, a proteolytic complex machinery regulating numerous cellular processes [70]. The chymotrypsin-like activity was the most sensitive towards SylA, although trypsin- and caspase-like activities were diminished as well. The crystal structure of yeast 20S proteasome with SylA revealed its covalent binding to the hydroxyl group of the active site Thr via a Michael-type 1,4-addition of the ThrO$^{\gamma}$ to the α, β-unsaturated amide moiety of the 12-membered ring system (Fig. 18) [70]. Those results strongly supported previous work by Coleman et al. [71] in which it was shown that SylA (**35**) exhibited a strong antiproliferative activity in human neuroblastoma and ovarian cancer cell lines in a dose-dependent manner, with an IC_{50} in the range of 20–25 μM. It also induced cell

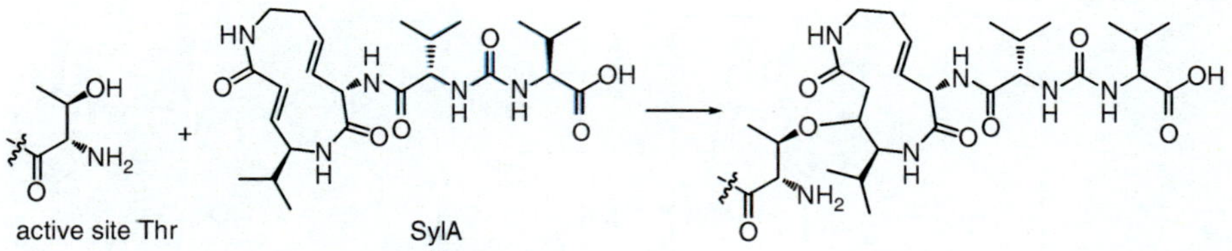

Fig. 18 Covalent inhibition mechanism of syringolin A

apoptosis; however, the cellular mechanism underlying those effects was not elucidated. Taken together, the results show that SylA exerts a promising anticancer activity due to the proteasome inhibition, thus expanding the repertoire of structural diversity of such therapeutic agents (i.e., bortezomib). Nevertheless, further development of a specific and effective anticancer agent requires in-depth investigation on side effects caused, for example, by cross-reactivity with other nucleophilic enzymes or reactivity in resistant cells due to rapid mutations of the proteasome active site. To answer these important questions, Clerc et al. [43] created SylA-based chemical probe **37** for ABPP. To facilitate an easy fluorescent read-out of cellular labeling events, a rhodamine tag was appended at the free carboxylic acid moiety of the dipeptide tail (Fig. 17). Rh-SylA (**37**) was examined in murine cell line EL4 and was able to efficiently label all β-subunits of both the constitutive proteasome and the immunoproteasome in a dose-dependent manner, with the strongest binding toward the β2 and β5 subunits. In contrast, bortezomib, a FDA-approved proteasome inhibitor used in cancer therapies, preferentially blocks β1 and β5 subunits. Interestingly, even at the concentration of 10 μM, which is far above cellular concentrations of anticancer drugs, no off-targets could be detected, although the reactive α,β-unsaturated carbonyl moiety could in principle also cross-react with more reactive cysteine proteases. This high specificity is presumably associated with the unusual structure of SylA and the accommodation of a side-chain cyclized inhibitor in S2 and S4 binding pockets. Also, **37** showed a significant labeling potency in clinically important leukemia cell lines, including bortezomib-adapted cells. Noticeably, active-site mutations originating from bortezomib treatment did not seem to influence preferential binding of SylA toward β2 and β5 subunits, thus creating an alternative for the next generation of potent proteasome inhibitors for the treatment of cancer. Most recently, the SylA-based ABPP probe was also used for fluorescence imaging and profiling of plant proteasome activity in vivo in *A. thaliana* [44], setting optimal conditions for studying proteasome activity in vivo in different organisms.

4.2.9 Showdomycin

The *C*-glycosyl nucleoside antibiotic showdomycin (**38**) was first isolated from *Streptomyces showdoensis* [72]. Its structure has striking similarity to uridine (**39**) and pseudouridine (**40**), but is distinguished by exhibiting a maleimide moiety (Fig. 19). The known high electrophilicity of this group towards *S*-nucleophiles

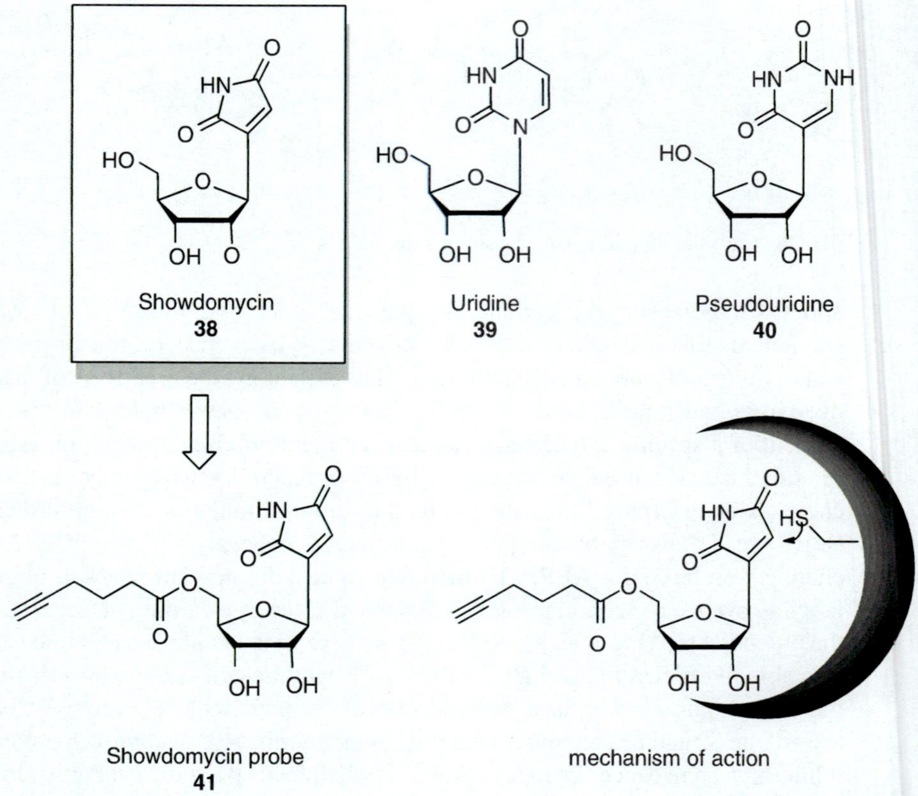

Fig. 19 Showdomycin and its mechanism of action

provoked early suggestions that this molecule exerts its biological function by covalent inhibition of active Cys residues of its protein targets. The structural similarity to uridine inspired, several years after its discovery the study of the effect of showdomycin on uridine metabolism. In this effort, the inhibitory effect of **38** against UMP kinase, uridine phosphorylase, and UDP-glucose dehydrogenase was reported and its mechanism of action by covalent binding confirmed [73].

More recently, Böttcher and Sieber undertook an ABPP approach for identifying the target proteins of **38** [74]. Recognizing the ideal situation of a built-in chemical warhead, they synthesized a showdomycin analog **41**, which featured an alkyne group suitable for click chemistry labeling. In an antibacterial assay, **38** and **41** showed similar values for minimum inhibitory concentrations (MIC) for *Staphylococcus aureus*, demonstrating that introduction of the label did not affect the antibiotic properties of the probe molecule. In another experiment, they compared the labeling pattern of the labeled showdomycin probe **41** with that of an N-labeled fluorescent maleimide probe. While the maleimide probe promiscuously labeled almost all proteins in the lysate, the showdomycin probe revealed selectivity for only a limited number of proteins. This important result substantiates the general view that, in protein-reactive natural products, the otherwise promiscuous reactive warhead is tamed by the sophisticated three-dimensional structures of these molecules and thus selectivity is gained. With **41** in hand, Böttcher and Sieber performed a comprehensive analysis of protein targets using the combined SDS-PAGE/MS identification approach, which revealed 13 different enzymes. These included enzymes that are crucial for oxidative stress resistance in *S. aureus*, such as alkylhydroperoxide reductases C (AhpC) and F (AhpF) and thioredoxin reductase (TrxB). The flexibility of their ABPP approach allowed them to compare the unique protein signature patterns of different strains of pathogenic and nonpathogenic bacteria. One important result from this investigation produced the insight that apparently the UDP-*N*-acetylglucosamine 1-carboxyvinyltransferases 1 (MurA1) and 2 (MurA2), which are involved in cell wall biosynthesis, are signature proteins for the MRSA strain Mu50. Although the expression levels of the corresponding genes *murA1* and *murA2* are comparable in all strains, the different amounts of active protein indicated that the difference in the in situ labeling is due to posttranslational regulation of enzyme activity [74].

4.2.10 Avrainvillamide

Under cellular conditions, the dimeric alkaloid stephacidin B was identified to be in equilibrium with its monomeric building block avrainvillamide (**42**), which is the true active species responsible for its known biological activity, namely the inhibition of tumor growth. Having mastered the total synthesis of both compounds, the Myers group attempted the annotation of the biological targets of **42** (Fig. 20) [75]. In their initial effort, they used a biotinylated activity-based probe **43** that incorporated a structurally less complex analog of avrainvillamide. Treatment of a LNCaP cell lysate with **43**, followed by affinity separation with streptavidin-agarose and SDS-PAGE, allowed the separation of several proteins that were addressed by **42**. MS identification showed that several Cys-containing proteins such as heat shock protein 60 (HSP60), exportin 1 (XPO1), glutathione reductase (GR), and peroxiredoxin 1 (PRX1) were protein binders of **42** [75]. In a subsequent study using the probe **44**, which is structurally more related to **42**, in a similar activity-based setup, they could demonstrate that avrainvillamide irreversibly binds to Cys275 of nucleophosmin, a known regulator of tumor suppressor protein p53 [76]. Depletion of nucleophosmin by RNA silencing led to increased sensitivity of HeLa cells towards apoptotic cell death in the presence of **42**, suggesting that the interaction of **42** with cellular nucleophosmin may play a role in the observed antiproliferative effects of this natural product [76].

4.2.11 Withaferin A

Extracts from the medicinal plant *Withania somnifera* have been used in Ayurvedic medicine for the treatment of chronic human diseases such arthritis and human

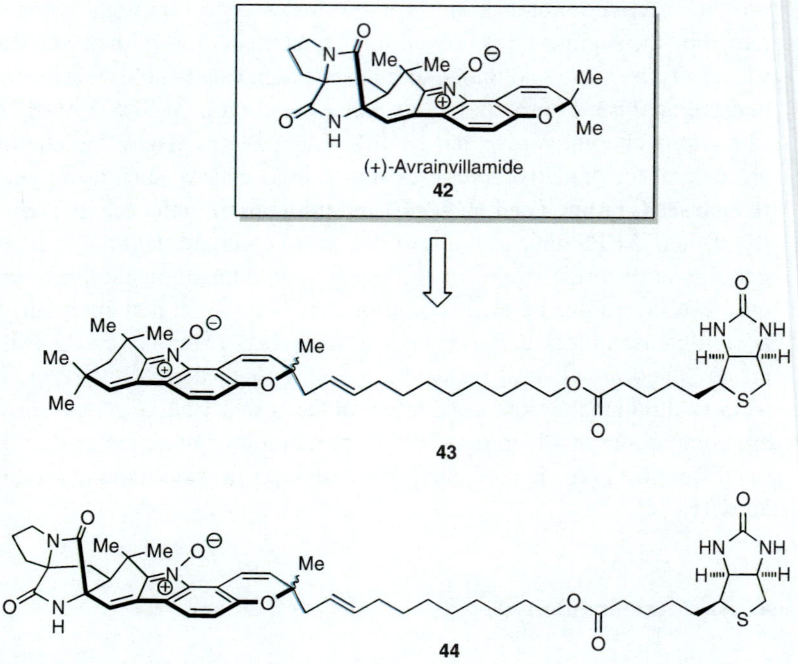

Fig. 20 (+)-Avrainvillamide and probe molecules

bleeding. Withaferin A (**45**) is a highly oxygenated steroid lactone found in this plant and has reported antitumor cytotoxic activities. Previous structure–activity studies, which indicated the importance of the 4β-hydroxy-5β,6β-epoxy-2-en-1-one moiety and the unsaturated lactone side chain for biological activity, guided Mohan et al. in preparation of the biotinylated analog **46**, which retained the biological activity of the parent compound **45** (Fig. 21) [77]. Affinity-based protein profiling identified 180 and 56 kDa proteins as covalent binding partners and suggested that another protein with 70 kDa might bind **45** reversibly [78]. In a subsequent publication about the nature of the 56 kDa protein, the group reported that it is the intermediate filament (IF) protein vimentin. **45** binds to this protein at Cys328 in the α-helical coiled coil 2B rod domain, probably via its A-ring electrophilic pharmacophore, consequently causing endothelial cell apoptosis [77].

4.3 Natural Products Containing a Lactone

4.3.1 Tetrahydrolipstatin

The β-lactone moiety represents a ubiquitous structural motif present in natural products and its moderately electrophilic heterocyclic ring reacts with nucleophilic residues of many proteins active sites. For example, esterastin is a known pancreatic

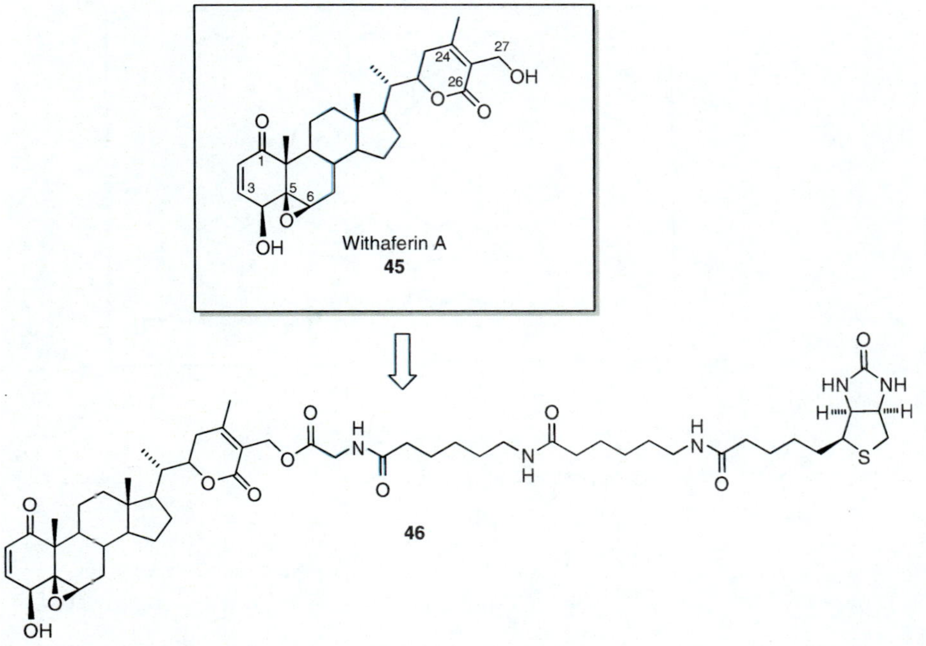

Fig. 21 Withaferin A and biotinylated derivative

esterase inhibitor produced by *Streptomyces lavendulae* [79], while its structurally related analog lipstatin (**47**) isolated from *Streptomyces toxytricini* [80] is a potent and selective inhibitor of pancreatic lipase. Other members belonging to this β-lactone family of inhibitors of pancreatic enzymes include ebelactones A and B [81], valilactone [82] and panclicins A–E [83]. On the other hand, hymeglusin [84] and obafluorin [85] are β-lactones displaying antifungal and antimicrobial activities. However, we still lack a comprehensive knowledge about the whole spectrum of cellular targets of β-lactones occurring in nature.

The most is known about lipstatin (**47**) and its close synthetic derivative tetrahydrolipstatin (THL, **48**) (Fig. 22). Lipstatin encompasses a β-lactone ring with two linear carbon chains. The longer carbon chain has two double bonds and an *N*-formylleucine ester side chain. THL (**48**) results from catalytic hydrogenation of unsaturated double bonds in lipstatin [86, 87]. Both compounds inhibit pancreatic lipase and other gastric lipases within the gastrointestinal tract by the nucleophilic attack of the active site Ser at the carbonyl carbon C1 of the β-lactone ring to form a β-hydroxyl serine ester (Fig. 23). THL is used as an anti-obesity drug under the trade name Orlistat to help obese patients lose weight by blocking dietary fat absorption [88, 89]. Interestingly, **48** was also recently found to inhibit the thioesterase domain of fatty acid synthase (FAS), an enzyme involved in the proliferation of cancer cells but not normal cells [90]. Orlistat (**48**) induces stress in tumor cells, blocks their growth and angiogenesis and, as a consequence, leads to a delay of tumor progression in various cancer types including prostate, breast,

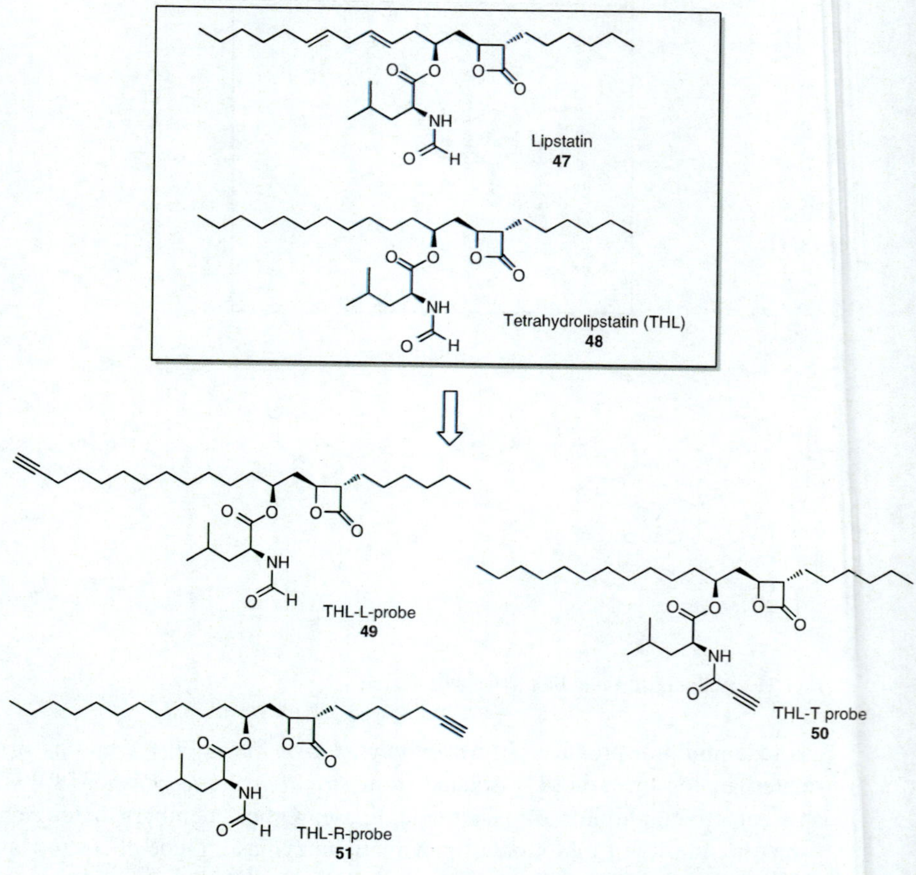

Fig. 22 Lipstatin and tetrahydrolipstatin

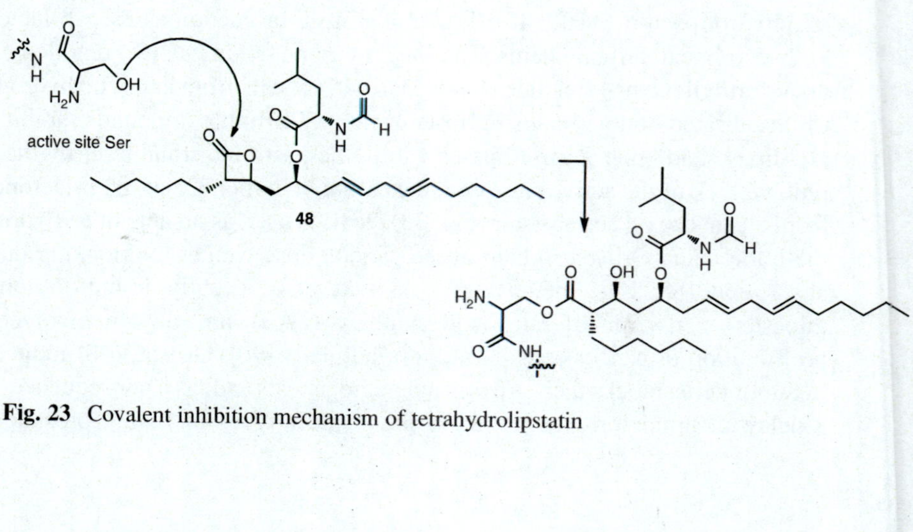

Fig. 23 Covalent inhibition mechanism of tetrahydrolipstatin

ovary, and melanoma. However, potential side effects of Orlistat, such as inhibition of other cellular off-targets or poor bioavailability, might hamper its application as an effective antitumor agent. To this end, the research group of Yao undertook an ABPP approach to look for new cellular targets of Orlistat, including its off-targets [91]. They prepared three THL-like chemical probes **49–51** (Fig. 22) by introducing very conservative modifications (alkyne tag) to the parent structure at different molecule locations. The modifications maintained its native properties but allowed target identification by conjugation of a reporter tag (rhodamine or biotin) via the bioorthogonal click chemistry reaction. First, all three probes were evaluated with respect to their ability to block cell proliferation, induce phosphorylation of the translation initiation factor eIF2α, and induce cell apoptosis by activation of caspase-8. These three cellular assays were able to compare the biological activity of the probes with that of the parent THL molecule. Collectively, the assays showed that introduction of the alkyne handle did not affect their biological activity, which was comparable with that of Orlistat. Next, all probes were incubated with live HepG2 cells (human hepatocellular carcinoma cell line) and, after downstream processing, the cellular targets were detected. Apart from the expected FAS enzyme, several other bands were also observed that were removed by treatment with an excess of Orlistat as competition. Enrichment of protein extracts, separation by SDS-PAGE, and MS/MS analysis led to the identification of eight new proteins, one of which was an unnamed protein. Two of the proteins, GAPDH (glyceraldehyde 3-phosphate dehydrogenase) and β-tubulin are housekeeping proteins expressed in most cells and responsible for cell division and energy production, respectively, but known to be expressed at elevated levels in tumor cells. Three other proteins, RLP7a, RLP14, and RPS9 are ribosomal proteins involved in protein synthesis and control of tumor growth, aggressiveness, and metastasis. The last two proteins, Annexin A2 and Hsp90AB1, play an important role in cell proliferation/division and protein degradation, respectively. The identity of all listed proteins labeled by the THL probes were also validated by western blot analysis with their respective antibodies, thus confirming that they are true cellular targets of **48**. This profiling study showcased unambiguously the importance of identification of other targets and elucidation of their role in connection to the potent drug candidate. In order to search for Orlistat derivatives of improved specificity towards FAS and decreased number of off-targets, another proteomic profiling was performed by the same research group [92]. They prepared a library of THL analogs diversified through a differently substituted triazole ring that was furnished by click chemistry reaction between alkyne-THL and various azides. Four derivatives with aromatic or sulfonamide residues displaying similar inhibitory activity towards FAS were identified; nevertheless, they also labeled most of the Orlistat off-targets, but to a lesser extent. Very recently [93], an expanded library built-up on Orlistat, encompassing modifications with regard to the position of the alkyne tag, identity of the incorporated amino ester, and chirality of a probe, was evaluated with the goal of improving the cellular activity of the probes. Strikingly, the authors could demonstrate that even very subtle changes in the structure caused a significant difference in cellular potency and target profile of the probes and allowed the identification of both common and unique protein targets. In this study, a cell-permeable, benzylguanine (BG)-containing Orlistat version was also implemented within an AGT/SNAP-tag strategy to deliver the probe to particular cell organelles in living cells with the aim of enhancing the cellular activity of Orlistat.

4.3.2 Lactacystin

Lactacystin (**52**) is a *Streptomyces* metabolite that inhibits cell cycle progression and induces neurite outgrowth in a murine neuroblastoma cell line. Although of low molecular weight, **52** embodies significant structural complexity exhibiting a [3.2.0] bicyclus containing a five-ring lactam and a four-ring lactone, and four consecutive stereogenic centers (Fig. 24). In order to investigate the origin of its interesting biological activity, Schreiber et al. prepared a tritium-labeled version of **52** by oxidizing the HO-group at the 9-position followed by reduction with tritiated sodium borohydride [94]. Incubation of crude extracts from Neuro-2a-cells or bovine brain with [^{3}H]lactacystin followed by SDS-PAGE and fluorography revealed the presence of an intensely labeled protein band of 24 kDa. Enrichment studies and Edman sequencing of tryptic fragments identified the 20S proteasome as its specific cellular target. The electrophilic lactone carbonyl of lactacystin (**52**) can covalently modify the highly conserved N-terminal Thr of the mammalian proteasome subunit X/MB1, thereby exhibiting considerable selectivity compared to other proteasome inhibitors [94].

4.3.3 Biomimetic β-Lactone Library

Recognizing the potential of β-lactones as a warhead for probe molecules, Böttcher and Sieber designed a small library of *trans*-β-lactones **53** (Fig. 25), which they screened against several bacterial strains [95]. Several targets of medicinal interest were identified, among them β-ketoacyl acyl carrier protein synthase II (KAS II), proline iminopeptidase (PIP), and caseinolytic protein protease (ClpP) [95]. ClpP, which is a phylogenetically highly conserved serine protease and an important virulence factor, warranted further investigation [96]. Chemical knockout treatment with

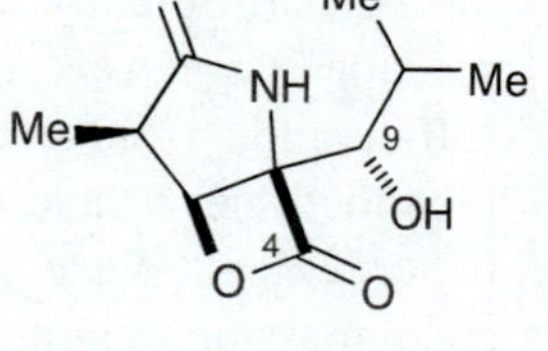

Lactacystin

52

Fig. 24 Lactacystin

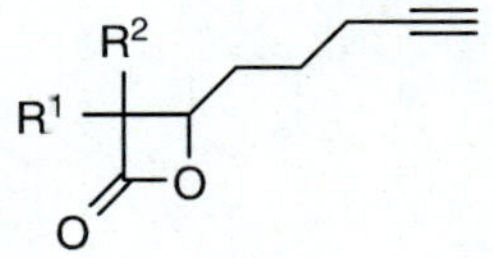

53

R^1 = H, Alkyl, Aryl
R^2 = H, Alkyl, Aryl

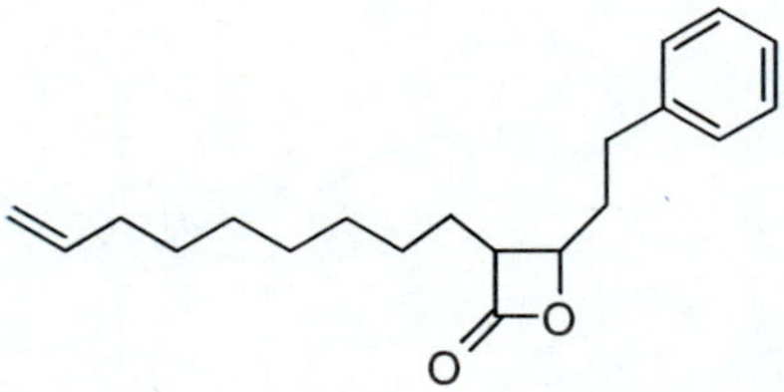

54

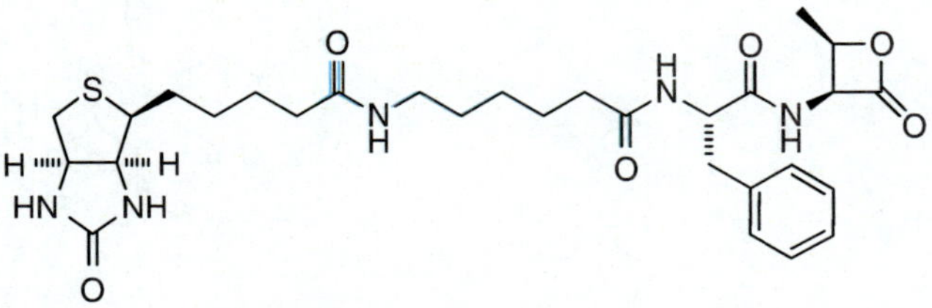

55

Fig. 25 β-Lactone library and tool compounds

improved ClpP inhibitor **54** resulted in significantly reduced secreted protein levels of important toxins [97], and downregulation of the expression of listeriolysin L (LLO) and phosphatidylinositol-specific phospholipase C (PI-PLC), which in consequence led to a decreased virulence of *Listeria monocytogenes* [98].

In a parallel effort, van der Hoorn et al. synthesized a library of biotinylated β-lactones for the ABPP monitoring of protease activity of plant extracts [99]. SDS-PAGE revealed several biotinylated compounds, from which MS analysis identified the 23 kDa band as protein P of the oxygen-evolving complex of photosystem II (PsbP). Remarkably, for this protein no nucleophilic Ser or Cys residues have been reported. Careful investigation of the labeling mechanism revealed that probe **55** reacts first with Cys-protease RD21, forming a thioester which then reacts with the N-terminus of PsbP. This impressive study teaches that initial ABPP results always need careful secondary screening and that further investigation of unexpected labeling sites can lead to intriguing molecular mechanisms [99].

4.4 Natural Products Containing a Lactam

4.4.1 Penicillin and β-Lactams

Penicillins are natural or semisynthetic antibiotics containing a reactive β-lactam functional group, which irreversibly reacts with proteins involved in cell wall biosynthesis. By definition, its target proteins are called penicillin-binding proteins

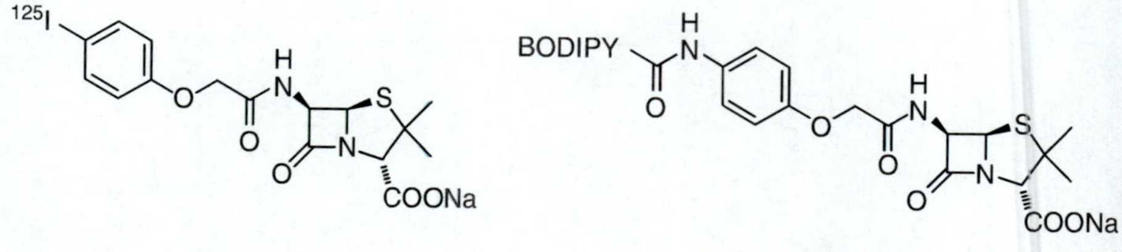

56 57

Fig. 26 Penicillin derivatives containing labels

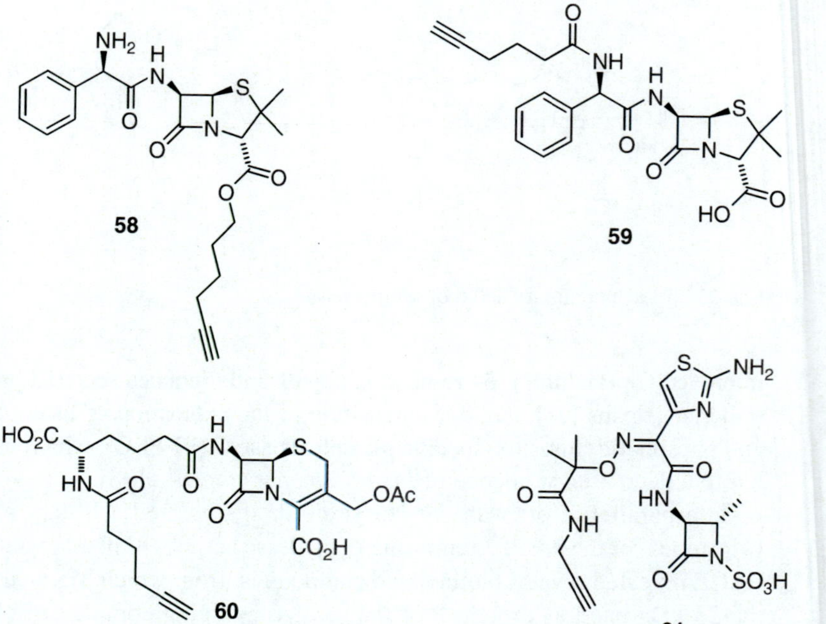

58 59 60 61

Fig. 27 β-Lactam activity-based probes inspired by commercial drugs

(PBPs). A group at Lilly has introduced the radiolabeled probe **56** and fluorescence-labeled probe **57** as tool compounds to profile the PBPs in different bacterial strains using SDS-PAGE (Fig. 26) [100, 101]. At this time, no MS analysis of the protein bands has been performed.

The group of Sieber has used an improved ABPP approach to look again at the targets of certain β-lactam antibiotics and prepared derivatives **58–61** by attaching an alkyne tag to clinically used parent compounds (Fig. 27). With these probes, they were able to perform in vivo experiments and identify a series of PBPs [102].

In a second line of research, they prepared a new set of synthetic β-lactam probes **62** around the monocyclic aztreonam structure (Fig. 28). Interestingly, none of these modified β-lactam probes labeled any PBPs but had preference for other enzymes such as β-ketoacyl acyl carrier protein III (KAS III), a β-lactamase,

Fig. 28 β-Lactam library

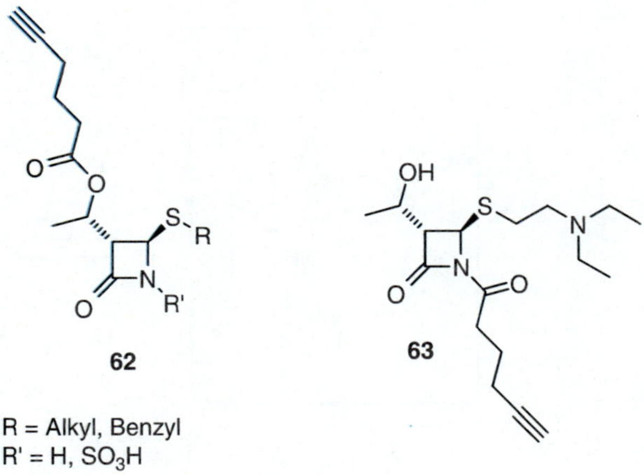

a lipase acylhydrolase, an antioxidant (AhpC), and the virulence-associated protein ClpP [102]. In a following publication, Staub and Sieber used one probe molecule **63** for the in situ profiling of antibiotic-sensitive and resistant *S. aureus* strains. With this probe they could identify five proteins unique for the MRSA strain. MS analysis revealed their identities as PBP2, PBP2′, the serine protease SPD_0, esterase E28, and a dipeptidase. These results offer not only a convenient way to monitor the activity of these enzymes important for the resistance of MRSA, but also a new lead structure for therapeutic intervention against these new protein targets [103].

4.5 Others

4.5.1 Acivicin

The natural product acivicin (**64**) is an amino acid containing a 3-chlorodihydroisoxazole ring. Its mode of action as an antibiotic is thought to occur via covalent modification of active site Cys or Thr of enzymes such as L-glutamine-dependent amidotransferases, glutamate synthase, and γ-glutamyl transpeptidase (γ-GT). Sieber et al. recognized the potential of the 3-halogenodihydroisoxazole warhead to serve as a scaffold for a small library of activity-based probe molecules [104]. Using efficient [3+2]-cycloaddition reactions they prepared an ensemble of regioisomeric compounds **65a** and **65b**, each containing an alkyne handle as a tag for their modification with rhodamine and biotin azides via Huisgen–azide–alkyne coupling (Fig. 29). Screening these probes against several different bacterial strains revealed the potential of these molecules. They identified pyrroline-5-carboxylase dehydrogenase, aldehyde dehydrogenase, bifunctional aldehyde-CoA/alcohol dehydrogenase, and 3-oxoacyl-[acyl-carrier-protein] synthase III (FabH) as targets covalently modified by their probes. Strikingly, the individual probes showed

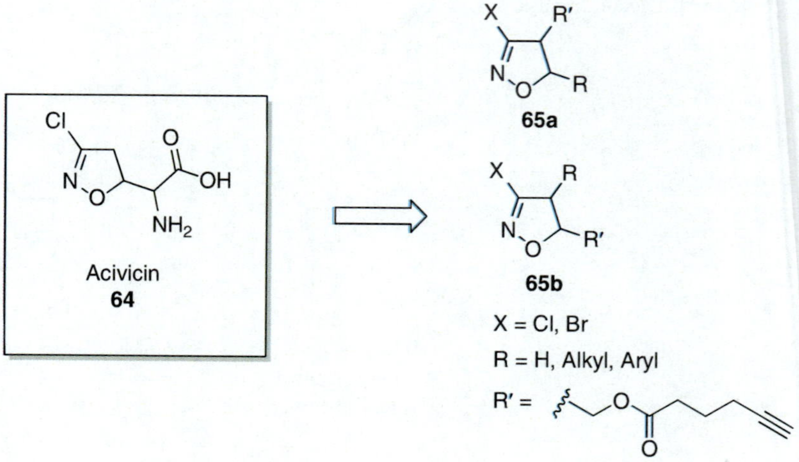

Fig. 29 Acivicin and a biomimetic library

considerable selectivity between these targets, and none of them labeled γ-GT, probably because of the missing amino acid recognition element. This study demonstrates that natural products serve as excellent starting points for probes against targets and can go beyond the target range of the original natural product.

5 Conclusions

In this review, we have collected several examples of natural products that react with their protein targets in a covalent manner. By introduction of an appropriate label onto these natural products, researchers were able to identify the target proteins responsible for the biological function. In the 1990s only one or two proteins could be annotated, but since then the impressive progress made in proteomic techniques, especially in MS instrumentation and data analysis, has opened a new dimension to more complete target annotation of natural products. We expect that ABPP will in the future be increasingly used in natural product research, and that the design of ABPP libraries based on natural product scaffolds, as pioneered by Cravatt and Sieber, will become a very important source of new probe molecules against new and biologically relevant targets.

Acknowledgements R.B. thanks the Austrian Science Fund (FWF) for a PhD scholarship for J.K. (Doktoratskolleg “Molecular Enzymology” W901-B05), and the PLACEBO (Platform for Chemical Biology)-project as part of the Austrian Genome Project GEN-AU funded by the Forschungsförderungsgesellschaft und Bundesministerium für Wissenschaft und Forschung for research funding.

References

1. Breinbauer R, Vetter IR, Waldmann H (2002) From protein domains to drug candidates - Natural products as guiding principles in the design and synthesis of compound libraries. Angew Chem Int Ed 41:2879–2890
2. Newman DJ, Cragg GM (2007) Natural products as sources of new drugs over the last 25 years. J Nat Prod 70:461–477
3. Clemons PA, Bodycombe NE, Carrinski HA, Wilson JA, Shamji AF, Wagner BK, Koehler AN, Schreiber SL (2010) Small molecules of different origins have distinct distributions of structural complexity that correlate with protein-binding profiles. Proc Natl Acad Sci USA 107:18787–18792
4. Carlson EE (2010) Natural products as chemical probes. ACS Chem Biol 5:639–653
5. Hopkins AL (2008) Network pharmacology: the next paradigm in drug discovery. Nat Chem Biol 4:682–690
6. Wong CC, Cheng KW, He Y-Y, Chen F (2008) Unraveling the molecular targets of natural products: Insights from genomic and proteomic analyses. Proteomics Clin Appl 2:338–354
7. Cheng KW, Wong CC, Wang M, He Q-Y, Chen F (2010) Identification and characterization of molecular targets of natural products by mass spectrometry. Mass Spectrom Rev 29:126–155
8. Dixon N, Wong LS, Geerlings TH, Micklefield J (2007) Cellular targets of natural products. Nat Prod Rep 24:1288–1310
9. Farha MA, Brown ED (2010) Chemical probes of *Escherichia coli* uncovered through chemical-chemical interaction profiling with compounds of known biological activity. Chem Biol 17:852–862
10. Sato S-I, Murata A, Shirakawa T, Uesugi M (2010) Biochemical target isolation for novices: affinity-based strategies. Chem Biol 17:616–623
11. Wang G, Shang L, Burgett AWG, Harran PG, Wang X (2007) Diazonamide toxins reveal an unexpected function for ornithine d-amino transferase in mitotic cell division. Proc Natl Acad Sci USA 104:2068–2073
12. Harding MW, Galat A, Uehling DE, Schreiber SL (1989) A receptor for the immunosuppressant FK506 is a cis-trans peptidyl-prolyl isomerise. Nature 341:758–760
13. Taunton J, Hassig CA, Schreiber SL (1996) A mammalian histone deacetylase related to the yeast transcriptional regulator Rpd3p. Science 272:408–411
14. Sato S-i, Murata A, Orihara T, Shirakawa T, Suenaga K, Kigoshi H, Uesugi M (2011) Marine natural product aurilide activates the OPA1-mediated apoptosis by binding to prohibitin. Chem Biol 18:131–139
15. Leslie BJ, Hergenrother PJ (2008) Identification of the cellular targets of bioactive small organic molecules using affinity reagents. Chem Soc Rev 37:1347–1360
16. Rix U, Superti-Furga G (2009) Target profiling of small molecules by chemical proteomics. Nat Chem Biol 5:616–624
17. Sadaghiani AM, Verhelst SHL, Bogyo M (2007) Tagging and detection strategies for activity-based proteomics. Curr Opin Chem Biol 11:20–28
18. Evans MJ, Cravatt BF (2006) Mechanism-based profiling of enzyme families. Chem Rev 106:3279–3301
19. Drahl C, Cravatt BF, Sorensen EJ (2005) Protein-reactive natural products. Angew Chem Int Ed 44:5788–5809
20. Koch MA, Schuffenhauer A, Scheck M, Wetzel S, Casaulta M, Odermatt A, Ertl P, Waldmann H (2005) Charting biologically relevant chemical space: a structural classification of natural products (SCONP). Proc Natl Acad Sci USA 102:17272–17277
21. Wilson RM, Danishefsky SJ (2006) Small molecule natural products in the discovery of therapeutic agents: the synthesis connection. J Org Chem 71:8329–8351
22. Böttcher T, Pitscheider M, Sieber SA (2010) Natural products and their biological targets: proteomic and metabolomic labelling strategies. Angew Chem Int Ed 49:2680–2698

23. Hamada K, Tamai M, Yamagishi M, Ohmura S, Sawada J, Tanaka I (1978) Isolation and characterization of E-64, a new thiol protease inhibitor. Agric Biol Chem 42:523–528
24. Matsumoto K, Mizoue K, Kitamura K, Tse WC, Huber CP, Ishida T (1999) Structural basis of inhibition of cysteine proteases by E-64 and its derivatives. Biopolymers 51:99–107
25. Barrett AJ, Kembhavi AA, Brown MA, Kirschke H, Knight CG, Tamai M, Hanada K (1982) L-trans-Epoxysuccinyl-leucylamido(4-guanidino)butane (E-64) and its analogues as inhibitors of cysteine proteinases including cathepsins B, H and L. Biochem J 201:189–198
26. Villadangos JA, Bryant RA, Deussing J, Driessen C, Lennon-Duménil AM, Riese RJ, Roth W, Saftig P, Shi GP, Chapman HA, Peters C, Ploegh HL (1999) Proteases involved in MHC class II antigen presentation. Immunol Rev 172:109–120
27. Gelb BD, Shi GP, Chapman HA, Desnick RJ (1996) Pycnodysostosis, a lysosomal disease caused by cathepsin K deficiency. Science 273:1236–1238
28. Beinfeld MC (1998) Prohormone and proneuropeptide processing. Recent progress and future challenges. Endocrine 8:1–5
29. Chapman HA, Reese RJ, Shi GP (1997) Emerging roles for cysteine proteases in human biology. Annu Rev Physiol 59:63–88
30. Iwata Y, Mort JS, Tateishi H, Lee ER (1997) Macrophage cathepsin L, a factor in the erosion of subchondral bone in rheumatoid arthritis. Arthritis Rheum 40:499–509
31. Yan S, Sameni M, Sloane BF (1998) Cathepsin B and human tumor progression. Biol Chem 379:113–123
32. Nomura T, Katunuma N (2005) Involvement of cathepsins in the invasion, metastasis and proliferation of the cancer cells. J Med Invest 50:1–9
33. Greenbaum D, Medzihradszky KF, Burlingame AL, Bogyo M (2000) Epoxide electrophiles as activity-dependent cysteine protease profiling and discovery tools. Chem Biol 7:569–581
34. Greenbaum D, Baruch A, Hayrapetian L, Darula Z, Burlingame A, Medzihradszky KF, Bogyo M (2002) Chemical approaches for functionally probing the proteome. Mol Cell Proteomics 1:60–68
35. Van der Hoorn RA, Leeuwenburgh MA, Bogyo M, Joosten MH, Peck SC (2004) Activity profiling of papain-like cysteine proteases in plants. Plant Physiol 135:1170–1178
36. Ahmed SU, Rojo E, Kovaleva V, Venkataraman S, Dombrowski JE, Matsuoka K, Raikhel NV (2000) The plant vacuolar sorting receptor AtELP is involved in transport of NH(2)-terminal propeptide-containing vacuolar proteins in Arabidopsis thaliana. J Cell Biol 149:1335–1344
37. Hayashi Y, Yamada K, Shimada T, Matsushima R, Nishizawa NK, Nishimura M, Hara-Nishimura I (2001) A proteinase-storing body that prepares for cell death or stresses in the epidermal cells of Arabidopsis. Plant Cell Physiol 42:894–899
38. Kaschani F, Verhelst SHL, van Swieten PF, Verdoes M, Wong C-S, Wang Z, Kaiser M, Overkleeft HS, Bogyo M, van der Hoorn RAL (2009) Minitags for small molecules: detecting targets of reactive small molecules in living plant tissues using 'click chemistry'. Plant J 57: 373–385
39. Hang HC, Loureiro J, Spooner E, van der Velden AW, Kim YM, Pollington AM, Maehr R, Starnbach MN, Ploegh HL (2006) Mechanism-based probe for the analysis of cathepsin cysteine proteases in living cells. ACS Chem Biol 1:713–723
40. Hanada M, Sugawara K, Kaneta K, Toda S, Nishiyama Y, Tomita K, Yamamoto H, Konishi M, Oki T (1992) Epoxomicin, a new antitumor agent of microbial origin. J Antibiot 45:1746–1752
41. Meng L, Mohan R, Kwok BHB, Elofsson M, Sin N, Crews CM (1999) Epoxomicin, a potent and selective proteasome inhibitor, exhibits in vivo antiinflammatory activity. Proc Natl Acad Sci USA 96:10403–10408
42. Sin N, Kim K, Elofsson M, Meng L, Auth H, Crews CM (1999) Total synthesis of the potent proteasome inhibitor epoxomicin: a useful tool for understanding proteasome biology. Bioorg Med Chem Lett 9:2283–2288
43. Clerc J, Florea BI, Kraus M, Groll M, Huber R, Bachmann AS, Dudler R, Driessen C, Overkleeft HS, Kaiser M (2009) Syringolin A selectively labels the 20 S proteasome in

murine EL4 and wild-type and bortezomib adapted leukaemic cell lines. ChemBioChem 10:2638–2643
44. Kolodziejek I, Misas-Villamil JC, Kaschani F, Clerc J, Gu C, Krahn D, Niessen S, Verdoes M, Willems LI, Overkleeft HS, Kaiser M, van der Hoorn RAL (2011) Proteasome activity imaging and profiling characterizes bacterial effector syringolin A. Plant Physiol 155:477–489
45. Florea BI, Verdoes M, Li N, van der Linden WA, Geurink PP, van den Elst H, Hofmann T, de Ru A, van Veelen PA, Tanaka K, Sasaki K, Murata S, den Dulk H, Brouwer J, Ossendorp FA, Kisselev AF, Overkleeft HS (2010) Activity-Based Profiling Reveals Reactivity of the Murine Thymoproteasome-Specific Subunit β5t. Chem Biol 17:795–801
46. Murata S, Sasaki K, Kishimoto T, Niwa S, Hayashi H, Takahama Y, Tanaka K (2007) Regulation of CD8+ T cell development by thymus-specific proteasomes. Science 316:1349–1353
47. Murata S, Takahama Y, Tanaka K (2008) Thymoproteasome: probable role in generating positively selecting peptides. Curr Opin Immunol 20:192–196
48. Leuenroth SJ, Okuhara D, Shotwell JD, Markowitz GS, Yu Z, Somlo S, Crews CM (2007) Triptolide is a traditional Chinese medicine-derived inhibitor of polycystic kidney disease. Proc Natl Acad Sci USA 104:4389–4394
49. McCallum C, Kwon S, Leavitt P, Shen D-M, Liu W, Gurnett A (2007) Triptolide binds covalently to a 90 kDa nuclear protein. Role of epoxides in binding and activity. Immunobiology 313:549–556
50. Titov DV, Gilman B, He Q-L, Bhat S, Low W-K, Dang Y, Smeaton M, Demain AL, Miller PS, Kugel JF, Goodrich JA, Liu JO (2011) XPB, a subunit of TFIIH, is a target of the natural product triptolide. Nat Chem Biol 7:182–188
51. Sin N, Meng L, Wang MQW, Wen JJ, Bornmann WG, Crews CM (1997) The anti-angiogenic agent fumagillin covalently binds and inhibits the methionine aminopeptidase, MetAP-2. Proc Natl Acad Sci USA 94:6099–6103
52. Evans MJ, Saghatelian A, Sorensen EJ, Cravatt BJ (2005) Target discovery in small-molecule cell-based screens by in situ proteome reactivity profiling. Nat Biotechnol 10:1303–1307
53. Wyman MP, Bulgarelli-Leva G, Zvelebil MJ, Pirola L, Vanhaesebroeck B, Waterfield MD, Panayotou G (1996) Wortmannin inactivates phosphoinositide 3-kinase by covalent modification of Lys-802, a residue involved in the phosphate transfer reaction. Mol Cell Biol 16:1722–1733
54. Yee M-C, Fas CF, Stohlmeyer MM, Wandless TJ, Cimprich KA (2005) A cell-permeable, activity base probe for protein and lipid kinases. J Biol Chem 280:29053–29059
55. Liu Y, Shreder KR, Gai W, Corral S, Ferris DK, Rosenblum JS (2005) Wortmannin, a widely used phosphoinositide 3-kinase inhibitor, also potently inhibits mammalian Polo-like kinase. Chem Biol 12:99–107
56. MacKintosh RW, Dalby KN, Campbell DG, Cohen PTW, Cohen P, MacKintosh C (1995) The cyanobacterial toxin microcystin binds covalently to cysteine-273 on protein phosphatase 1. FEBS Lett 371:236–240
57. Runnegar M, Berndt N, Kong S-M, Lee EYC, Zhang L (1995) In vivo and in vitro binding of microcystin to protein phosphatase 1 and 2A. Biochem Biophys Res Commun 216:162–169
58. Shreder KR, Liu Y, Nomanhboy T, Fuller SR, Wong MS, Gai WZ, Wu J, Leventhal PS, Lill JR, Corral S (2004) Design and synthesis of AX7574: a microcystin-derived, fluorescent probe for serine/threonine phosphatases. Bioconjug Chem 15:790–798
59. Kwok BHB, Koh B, Ndubuisi MI, Elofsson M, Crews CM (2001) The anti-inflammatory natural product parthonlide from the medicinal herb ferverfew directly binds to and inhibits IκB kinase. Chem Biol 8:759–766
60. Kunzmann MH, Straub I, Böttcher T, Sieber SA (2011) Protein reactivity of natural product-derived γ-butyrolactones. Biochemistry 50:910–916
61. Teruya T, Simizu S, Kanoh N, Osada H (2005) Phoslactomycin targets cysteine-269 of the protein phosphatise 2A catalytic subunit in cells. FEBS Lett 579:2463–2468

62. Kanoh N, Takayama H, Honda K, Moriya T, Teruya T, Simizu S, Osada H, Iwabuchi Y (2010) Cleavable linker for photo-cross-linked small-molecule affinity matrix. Bioconjug Chem 21:182–186
63. Usui T, Watanabe H, Nakayama H, Tada Y, Kanoh N, Kondoh M, Asao T, Takio K, Watanabe H, Nishikawa K, Kitahara T, Osada H (2004) The anticancer natural product pironetin selectively targets Lys352 of α-tubulin. Chem Biol 11:799–806
64. Kudo N, Wolff B, Sekimoto T, Schreiner EP, Yoneda Y, Yanagida M, Horinouchi S, Yoshida M (1998) Leptoymcin B inhibition of signal-mediated nuclear export by direct binding to CRM1. Exp Cell Res 242:540–547
65. Kudo N, Matsumori N, Taoka H, Fujiwara D, Schreiner EP, Wolff B, Yoshida M, Horinouchi S (1999) Leptomycin B inactivates CRM17exportin 1 by covalent modification at a cysteine residue in the central conserved region. Proc Natl Acad Sci USA 96:9112–9117
66. Vanderwal CD, Vosburg DA, Weiler S, Sorensen EJ (2003) An enantioselective synthesis of FR182877 provides a chemical rationalization of its structure and affords multigram quantities of its direct precursor. J Am Chem Soc 125:5393–5407
67. Adam GC, Vanderwal CD, Sorensen EJ, Cravatt BF (2003) (−)-FR182877 is a potent and selective inhibitor of carboxylesterase-1. Angew Chem Int Ed 42:5480–5484
68. Buey RM, Calvo E, Barasoain I, Pineda O, Edler MC, Matesanz R, Cerezo G, Vanderwal CD, Day BW, Sorensen EJ, Lopez JA, Andreu JM, Hamel E, Diaz JF (2007) Cyclostreptin binds covalently to microtubule pores and lumenal taxoid binding sites. Nat Chem Biol 3:117–125
69. Wäspi U, Blanc D, Winkler T, Ruedi P, Dudler R (1998) Syringolin, a novel peptide elicitor from Pseudomonas syringae pv. syringae that induces resistance to *Pyricularia oryzae* in rice. Mol Plant Microbe Interact 11:727–733
70. Groll M, Schellenberg B, Bachmann AS, Archer CR, Huber R, Powell TK, Lindow S, Kaiser M, Dudler R (2008) A plant pathogen virulence factor inhibits the eukaryotic proteasome by a novel mechanism. Nature 452:755–758
71. Coleman CS, Rocetes JP, Park DJ, Wallick CJ, Warn-Cramer BJ, Michel K, Dudler R, Bachmann AS (2006) Syringolin A, a new plant elicitor from the phytopathogenic bacterium Pseudomonas syringae pv. syringae, inhibits the proliferation of neuroblastoma and ovarian cancer cells and induces apoptosis. Cell Prolif 39:599–609
72. Nishimura H, Mayama M, Komatsu Y, Kato H, Shimaoka N, Tanaka Y (1964) Showdomycin, a new antibiotic from a streptomyces sp. J Antibiot 17:148–155
73. Roy-Burman S, Roy-Burman P, Visser DW (1968) Showdomycin, a new nucleoside antibiotic. Cancer Res 28:1605–1610
74. Böttcher T, Sieber SA (2010) Showdomycin as a versatile chemical tool for the detection of pathogenesis-associated enzymes in bacteria. J Am Chem Soc 132:6964–6972
75. Wulff JE, Herzon SB, Siegrist R, Myers AG (2007) Evidence for the rapid conversion of stephacidin B into the electrophilic monomer avrainvillamide in cell culture. J Am Chem Soc 129:4898–4899
76. Wulff JE, Siegrist R, Myers AG (2007) The natural product avrainvillamide binds to the oncoprotein nucleophosmin. J Am Chem Soc 129:14444–14451
77. Bargagna-Mohan P, Hamza A, Y-e K, Ho YK, Mor-Vaknin N, Wendschlag N, Liu J, Evans RM, Markovitz DM, Zhan C-G, Kim KB, Mohan R (2007) The tumor inhibitor and antiangiogenic agent withaferin A targets the intermediate filament protein vimentin. Chem Biol 14:623–634
78. Yokota Y, Bargagna-Mohan P, Ravindranath PP, Kim KB, Mohan R (2006) Development of withaferin A analogs as probes angiogenesis. Bioorg Med Chem Lett 16:2602–2607
79. Umezawa H, Aoyagi T, Hazato T, Uotani K, Kojima F, Hamada M, Takeuchi T (1978) Esterastin, an inhibitor of esterase, produced by Actinomycetes. J Antibiot 31:639–641
80. Weibel EK, Hadvary P, Hochuli E, Kupfer E, Lengsfeld H (1987) Lipstatin, an inhibitor of pancreatic lipase, produced by Streptomyces toxytricini. I. Producing organism, fermentation, isolation and biological activity. J Antibiot 40:1081–1085

81. Umezawa H, Aoyagi T, Uotani K, Hamada M, Takeuchi T, Takahashi S (1980) Ebelactone, an inhibitor of esterase, produced by actinomycetes. J Antibiot 33:1594–1596
82. Kitahara M, Asano M, Naganawa H, Maeda K, Hamada M, Aoyagi T, Umezawa H, Iitaka Y, Nakamura H (1987) Valilactone, an inhibitor of esterase, produced by actinomycetes. J Antibiot 35:1647–1650
83. Mutoh M, Nakada N, Matsukuma S, Ohshima S, Yoshinari K, Watanabe J, Arisawa M (1994) Panclicins, novel pancreatic lipase inhibitors I. Taxonomy, fermentation, isolation and biological activity. J Antibiot 47:1369–1375
84. Omura S, Tomoda H, Kumagai H, Greenspan MD, Yodkovitz JB, Chen JS, Alberts AW, Martin I, Mochales S, Monaghan RL, Chabala JC, Schwartz RE (1987) Potent inhibitory effect of antibiotic 1233a on cholesterol biosynthesis which specifically blocks 3-hydroxy-3-methylglutaryl coenzyme A synthase. J Antibiot 40:1356–1357
85. Wells JS, Trejo WH, Principe PA, Sykes RB (1984) Obafluorin, a novel β-lactone produced by Pseudomonas fluorescens. Taxonomy, fermentation and biological properties. J Antibiot 37:802–803
86. Barbier P, Schneider F (1987) Syntheses of tetrahydrolipstatin and absolute configuration of tetrahydrolipstatin and lipstatin. Helv Chim Acta 70:196–202
87. Barbier P, Schneider F, Widmer U (1987) Stereoselective Syntheses of tetrahydrolipstatin and of an analogue, potent pancreatic-lipase inhibitors containing a β-lactone moiety. Helv Chim Acta 70:1412–1418
88. Heck AM, Yanovski JA, Calis KA (2000) Orlistat, a new lipase inhibitor for the management of obesity. Pharmacotherapy 20:270–279
89. Lucas KH, Kaplan-Machlis B (2001) Orlistat – a novel weight loss therapy. Ann Pharmacother 35:314–328
90. Kridel SJ, Axelrod F, Rozenkrantz N, Smith JW (2004) Orlistat is a novel inhibitor of fatty acid synthase with antitumor activity. Cancer Res 64:2070–2075
91. Yang P-Y, Liu K, Ngai MH, Lear MJ, Wenk MR, Yao SY (2010) Activity-based proteome profiling of potential cellular targets of orlistat – an FDA approved drug with anti-tumor activities. J Am Chem Soc 132:656–666
92. Ngai MH, Yang P-Y, Liu K, Shen Y, Wenk MR, Yao SY (2010) Click-based synthesis and proteomic profiling of lipstatin analogues. Chem Commun 46:8335–8337
93. Yang P-Y, Liu K, Zhang C, Chen GYJ, Shen Y, Ngai MH, Lear MJ, Yao SY (2011) Chemical Modification and Organelle-Specific Localization of Orlistat-Like Natural-Product-Based Probes. Chem Asian J 6:2762–2775
94. Fenteany G, Standaert RF, Lane WS, Choi S, Corey EJ, Schreiber SL (1995) Inhibition of proteasome activites and subunit-specific amino-terminal threonine modification by lactacaystin. Science 268:726–731
95. Böttcher T, Sieber SA (2008) β-Lactones as privileged structures for the active-site labelling of versatile bacterial enzyme classes. Angew Chem Int Ed 47:4600–4603
96. Böttcher T, Sieber SA (2008) β-Lactones as specific inhibitors of ClpP attenuate the production of extracellular virulence factors of Staphylococcus aureus. J Am Chem Soc 130:14400–14401
97. Böttcher T, Sieber SA (2009) Structurally refined β-lactones as potent inhibitors of devastating bacterial virulence factors. ChemBioChem 10:663–666
98. Böttcher T, Sieber SA (2009) β-lactones decrease the intracellular virulence of *Listeria monocytogenes* in marcophages. ChemMedChem 4:1260–1263
99. Wang Z, Gu C, Colby T, Shindo T, Balamurugan R, Waldmann H, Kaiser M, van der Hoorn RAL (2008) β-Lactone probes identify a papain-like peptide ligase in *Arabidopsis thaliana*. Nat Chem Biol 4:557–563
100. Preston DA, Wu CYE, Blaszczak LC, Seitz DE, Halligan NG (1990) Biological characterization of a new radioactive labelling reagent for bacterial penicillin-binding proteins. Antimicrob Agents Chemother 34:718–721

101. Zhao G, Meier TI, Kahl SD, Gee KR, Blaszczak LC (1999) BOCILLIN FL, a sensitive and commercially available reagent for detection of penicillin-binding proteins. Antimicrob Agents Chemother 43:1124–1128
102. Staub I, Sieber SA (2008) β-Lactams as selective chemical probes for the in vivo labelling of bacterial enzymes involved in cell wall biosynthesis, antibiotic resistance, and virulence. J Am Chem Soc 130:13400–13409
103. Staub I, Sieber SA (2009) β-Lactam probes as selective chemical-proteomic tools for the identification and functional characterization of resistance associated enzymes in MRSA. J Am Chem Soc 131:6271–6276
104. Orth R, Böttcher T, Sieber SA (2010) The biological targets of acivicin inspired 3-chlro- and 3-bromodihydroisoxazole scaffolds. Chem Commun 46:8475–8477

Top Curr Chem (2012) 324: 85–114
DOI: 10.1007/128_2011_286
© Springer-Verlag Berlin Heidelberg 2011
Published online: 26 October 2011

Photoaffinity Labeling in Activity-Based Protein Profiling

Paul P. Geurink, Laurette M. Prely, Gijs A. van der Marel, Rainer Bischoff, and Herman S. Overkleeft

Abstract Activity-based protein profiling has come to the fore in recent years as a powerful strategy for studying enzyme activities in their natural surroundings. Substrate analogs that bind covalently and irreversibly to an enzyme active site and that are equipped with an identification or affinity tag can be used to unearth new enzyme activities, to establish whether and at what subcellular location the enzymes are active, and to study the inhibitory effects of small compounds. A specific class of activity-based protein probes includes those that employ a photo-activatable group to create the covalent bond. Such probes are targeted to those enzymes that do not employ a catalytic nucleophile that is part of the polypeptide backbone. An overview of the various photo-activatable groups that are available to chemical biology researchers is presented, with a focus on their (photo)chemistry and their application in various research fields. A number of comparative studies are described in which the efficiency of various photo-activatable groups are compared.

Keywords Activity-based profiling · Chemical biology · Photoaffinity labeling · Photocrosslinking · Proteomics

P.P. Geurink, G.A. van der Marel, and H.S. Overkleeft (✉)
Leiden Institute of Chemistry and the Netherlands Proteomics Centre, Leiden University, Einsteinweg 55, 9502, 2300 Leiden, The Netherlands
e-mail h.s.overkleeft@chem.leidenuniv.nl

L.M. Prely, and R. Bischoff
Department of Analytical Biochemistry, University Centre for Pharmacy, University of Groningen, Antonius Deusinglaan 1, 9713 Groningen, The Netherlands

Contents

Abbreviations

ABP	Activity-based probe
ABPP	Activity-based protein profiling
ADAM	A disintegrin and metalloproteinase
A*f*BP	Affinity-based probe
HDAC	Histone deacetylase
Lck	Leukocyte-specific protein tyrosine kinase
MMP	Matrix metalloproteinase
MTase	Methyltransferase
PAL	Photoaffinity labeling
PS1	Presenilin 1
SAH	*S*-Adenosyl-L-homocysteine
SAM	*S*-Adenosyl-L-methionine
TIP	Target identification probe
Tmd	3-(Trifluoromethyl)-3*H*-diazirine

1 Introduction

Proteomics research focuses on the study of proteins, their functioning, and their interactions with other biomolecules in the context of complex biological samples. Rather than focusing on a single protein, proteomics research takes on large numbers of distinct proteins at the same time, ideally the whole pool of proteins (called the proteome) expressed at a given time by an organism or cell type [1, 2]. In the global study of biological systems, proteomics is situated between genomics (study of the genome) and metabolomics (study of metabolites produced in cellular processes). The ultimate goal in proteomics is the complete understanding of each single protein in all physiological processes, which is of great importance in understanding pathological states. Since it is extremely difficult to study all proteins and their properties at once, proteomic research is usually performed by taking on one specific subset of proteins from two or more different biological systems (e.g., healthy cells and infected cells) in a comparative study. Traditionally, this is done by separation of the protein subset of choice, for example by one- or two-dimensional gel electrophoresis and immunoprecipitation, followed by determination of abundance levels.

An attractive alternative strategy developed for proteomics research is activity-based protein profiling (ABPP) [3, 4]. This strategy is based on the determination of an enzyme's activity rather than its levels of expression, which is important since an enzyme's abundance is not necessarily linked to its activity in biological processes. A reason for this is that a large number of enzymes are expressed as inactive proenzymes and therefore need to be transformed into their active form. In addition, the enzyme's state can be switched between active and inactive by posttranslational modifications such as phosphorylation, glycosylation, acetylation, ubiquitylation, and methylation or by their natural inhibitors. This chapter focuses on ABPP studies, with specific focus on the implementation of photo-activatable probes. In Sect. 2, some general aspects of ABPP will be discussed. Section 3 deals with the various photophores that are used to profile enzymes or enzyme families. Attention is given to their preparation, their photochemistry, their application in chemical biology studies, the pros and cons of the various photoreactive groups, and in some cases their comparative efficiency in selected studies.

2 Activity-Based and Affinity-Based Protein Profiling

ABPP makes use of relatively small organic molecules to label a specific enzyme (family) in its active state. These organic molecules are called "activity-based probes" (ABPs) and comprise three major elements. The recognition unit directs the ABP to the target enzyme and is designed to resemble structural and functional motives of the natural substrate of the target enzyme. This moiety is often designed such that a number of related enzymes within a family is targeted. Closely attached to the recognition part is a reactive group (also termed "warhead"), which reacts in the enzyme's active site and thereby establishes a covalent bond between the ABP and enzyme. Depending on the type of enzyme and reactive group, this interaction can be either reversible or irreversible. The recognition element and the warhead are linked, most often via a spacer, to a reporter group or "tag", which allows visualization and/or purification of the bound enzyme. Commonly used tags are fluorophores, radioactive labels, epitope tags, and biotin, of which the latter can be used for both visualization and purification purposes (Fig. 1, top).

An elegant extension to ABPP has been developed for those cases in which the reporter group hampers the interaction between ABP and enzyme or drastically lowers the ability to cross the cell's membrane, which is of interest especially for labeling in living cells. In this approach (referred to as two-step labeling) the ABP's tag is replaced by a ligation handle, which can be connected to the reporter group after the enzyme has been captured (Fig. 1, bottom) [5]. A requirement for this ligation handle is that it is unreactive towards all functionalities present in a

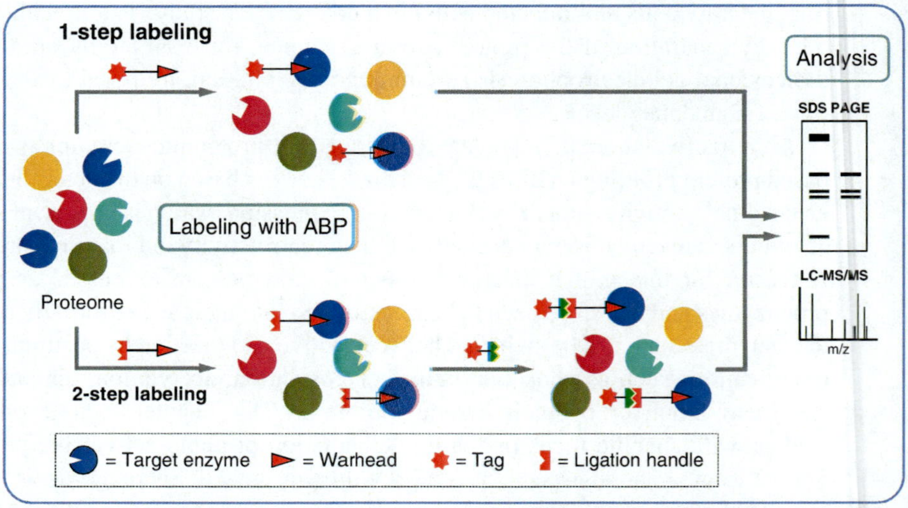

Fig. 1 An ABPP experiment. Both the one-step (*top*) and the two-step (*bottom*) labeling strategies are shown

biological sample (bioorthogonal). The most popular ligation handles are the azide and the (terminal) alkyne. The ligation reactions used (shown in Fig. 2) are the Staudinger–Bertozzi ligation [6] and the Huisgen 2,3-dipolar cycloaddition or "click" reaction, which can be divided into copper-catalyzed [7, 8] and strain-promoted reactions [9, 10]. Recently, Boons and coworkers developed a new reagent, which allows for a strain-promoted click reaction after irradiation with light and is therefore termed the "photoclick" reaction [11]. In addition, the use of the Diels–Alder reaction as alternative ligation reaction is subject to growing attention [12] and of the different Diels–Alder type reactions, the inverse-electron-demand variation appears the most promising for in vivo labeling [13, 14].

The ABPP strategy as shown in Fig. 1 is especially well suited for those enzymes that contain a nucleophilic amino acid side chain residue in their active site (e.g., serine, cysteine, threonine), which is responsible for the enzyme's catalytic activity. The ABP reactive group, which binds the target enzyme, is designed such that it reacts with this nucleophilic residue to form a covalent bond and is therefore named "electrophilic trap". Examples of enzymes targeted with this strategy are cysteine proteases [15–18], serine hydrolases [19–22], and proteasome subunits [23–25]. A difficulty arises for enzymes that do not rely on a nucleophilic residue in their active site, which precludes the use of an electrophilic trap. Among these are the metalloproteases, histone deacetylases (HDACs) (both of which employ a water molecule for their catalytic activity), and kinases. A useful alternative for the use of an electrophilic trap is the so-called photoaffinity labeling (PAL), in which the probe used is commonly referred to as "affinity-based probe" (A*f*BP) [26]. The basic principle is shown in Fig. 3. In this approach the A*f*BP binds the target enzyme in a reversible manner, either via noncovalent interactions (electrostatic, hydrophobic) or via a reversible covalent bond. Although these interactions can be

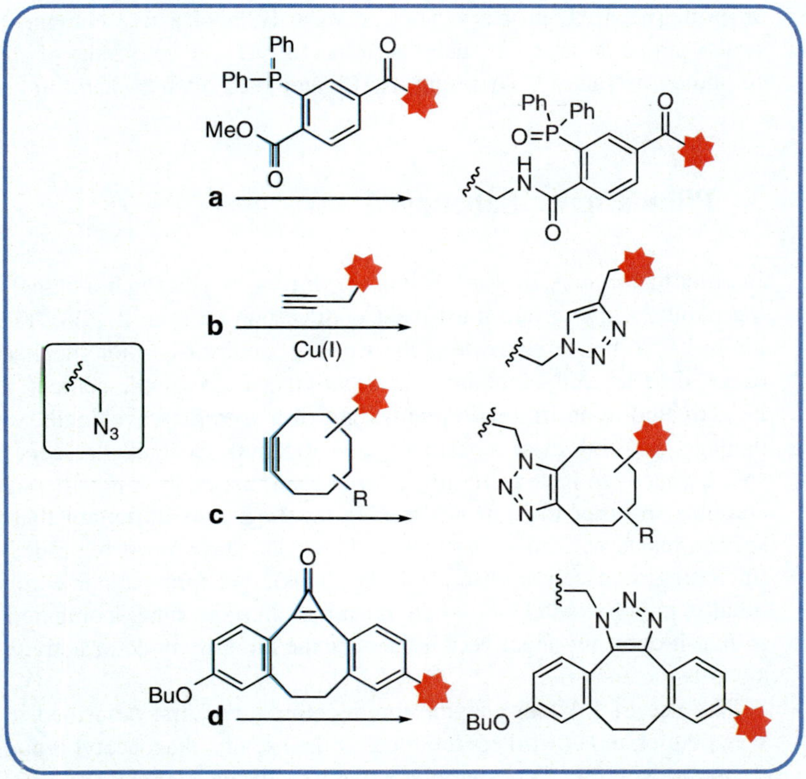

Fig. 2 Ligation reactions used in two-step ABPP using the azide functionality as ligation handle: (**a**) Staudinger–Bertozzi ligation, (**b**) Cu(I)-catalyzed click reaction, (**c**) copper-free click reaction, and (**d**) photoclick reaction

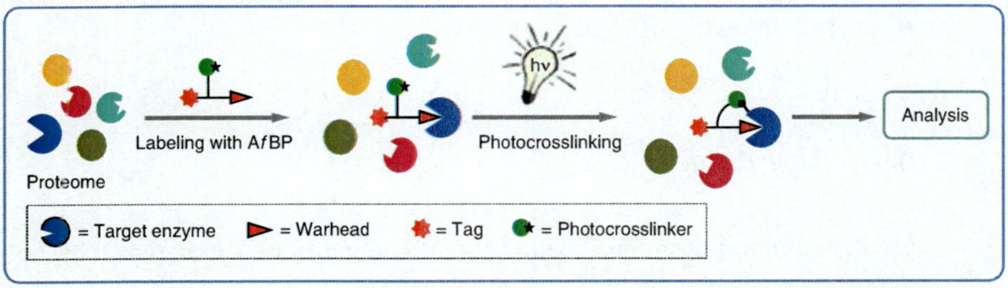

Fig. 3 Basic principle of photoaffinity labeling (PAL), which requires an affinity-based probe (A*f*BP) and UV irradiation

relatively strong, they cannot withstand harsh denaturing conditions often applied in biochemical protocols. An additional feature of the A*f*BP is the introduction of a photoreactive group (also termed photophore or photocrosslinker), which forms an irreversible covalent bond between probe and enzyme upon activation by light.

In principle, PAL probes do not necessarily label active enzymes, however the probe can be designed as such that it has to enter the active site of an enzyme prior to photocrosslinking. Therefore, ABPP and PAL often go hand-in-hand.

3 Photoaffinity Labeling

In order to use an A*f*BP for labeling of (active) enzymes in biological environments the photoreactive group must meet with certain criteria [27, 28]. First, the photophore has to be stable towards the various conditions a biological sample may be exposed to, as well as the intrinsic reactivity of the sample content, and must only be activated upon irradiation with light of a specific wavelength, which may not damage the biological system ($\lambda_{act} > 300$ nm). Second, the generated reactive species needs to have a lifetime shorter than that of the studied enzyme–substrate complex in order to limit nonspecific labeling. It is important that the activated species reacts with any chemical entity in close proximity, regardless of its nature (including relatively unreactive C–H bonds), and forms a stable covalent adduct. Finally, the photoreactive moiety must be relatively small, compared to the probe, so that it does not negatively influences the binding mode or activity of the A*f*BP towards the enzyme.

The use of PAL in enzyme modifications was first described in 1962, when Westheimer and coworkers reported on the use of a diazoacetyl group to inactivate chymotrypsin [29]. Considerable research on the development of new PAL reagents has taken place ever since [27, 28, 30–36], but only a few photophores, which largely meet the above-mentioned requirements, are being used nowadays in A*f*BPs. These are aryl azides (first reported use in 1969) [37], diazirines (1973) [38] and benzophenones (1973) [39]. The chemistry of these three photoreactive groups after photolysis as well as their use in recently reported A*f*BPs will be discussed in Sects. 3.1–3.5.

3.1 Aryl Azide

Upon activation of an aryl azide (**1**, see Scheme 1) by irradiation with light of the appropriate wavelength, molecular nitrogen (N_2) is expelled and a singlet nitrene (**2**) is formed initially. This highly energetic, highly reactive species has a short lifetime (~0.1 ms) and is quickly converted into other intermediates [35, 40]. Intersystem crossing leads to a triplet nitrene (**3**), which is about 20 kcal/mol lower in energy [41]. A major difference between the two nitrene states is their nature of reactivity. Singlet nitrenes behave like electrophiles and can readily undergo an insertion reaction with C–H bonds (giving compound **4**), whereas the triplet state can be seen as a diradical, which first abstracts a hydrogen radical from a nearby C–H (**9**) followed by coupling to the formed carbon radical. Although they react via two

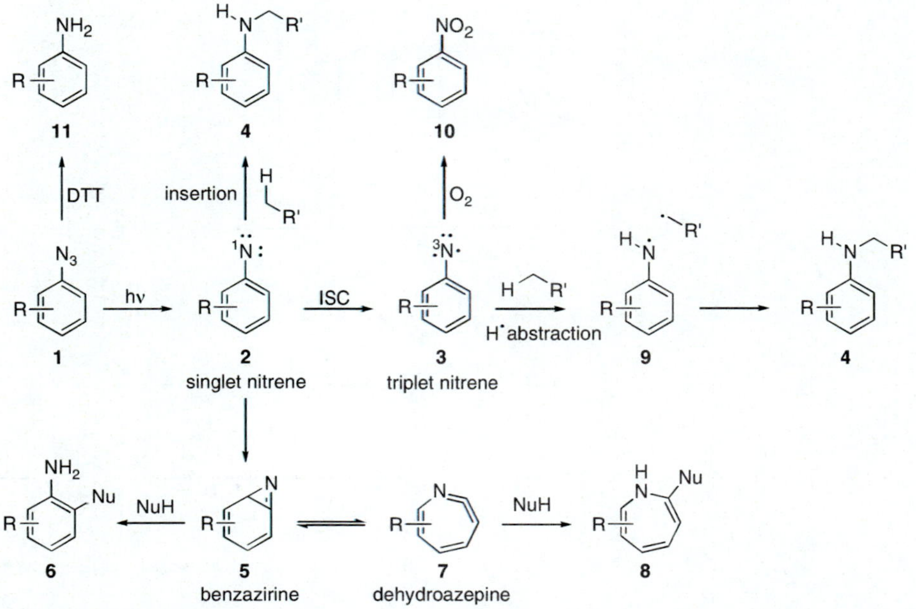

Scheme 1 Possible reaction mechanisms of the reactive intermediates formed after photolysis of aryl azides. *DTT* dithiothreitol, *ISC* intersystem crossing, *NuH* nucleophile

different mechanisms, the product is the same (**4**). Singlet nitrenes can also undergo a rapid rearrangement into the corresponding benzazirine (**5**), which can further rearrange into dehydroazepine (**7**). Both these species are long-lived electrophiles and can react with a nearby nucleophile, which results in compounds **6** and **8**, respectively. Two observed side-reactions that are not to be ignored when the aryl azide is applied in PAL are aerobic oxidation of the triplet nitrene to the corresponding nitro species **10** [42] and reduction of the initial aryl azide **1** to the amine **11** by dithiols such as dithiothreitol [43].

Aryl azides (such as **13**) can be easily prepared from their corresponding amines (**12**) in one or two steps. Three examples are given in Scheme 2a. The most common method is the diazotization of the amine with sodium nitrite under acidic conditions (giving diazonium species **14**), followed by addition of sodium azide in an aqueous medium (route *i* in Scheme 2a) [44]. In 2003, the synthetic method was improved by application of triflyl azide (TfN_3), which allowed a one-step conversion and gave improved yields (route *ii* in Scheme 2a) [45]. Recently, the development of sulfonyl azides **15** (with R = imidazole [46] or R = benzotriazole [47]) was reported; these proved to be more stable reagents and allowed a conversion under mild conditions (route *iii* in Scheme 2a).

A major drawback of aryl azides is that their maximum absorption wavelength is below 300 nm, since electromagnetic irradiation at these wavelengths can substantially damage the biological system. Consequently, a large number of substituted aryl azides have been made and evaluated for their absorption

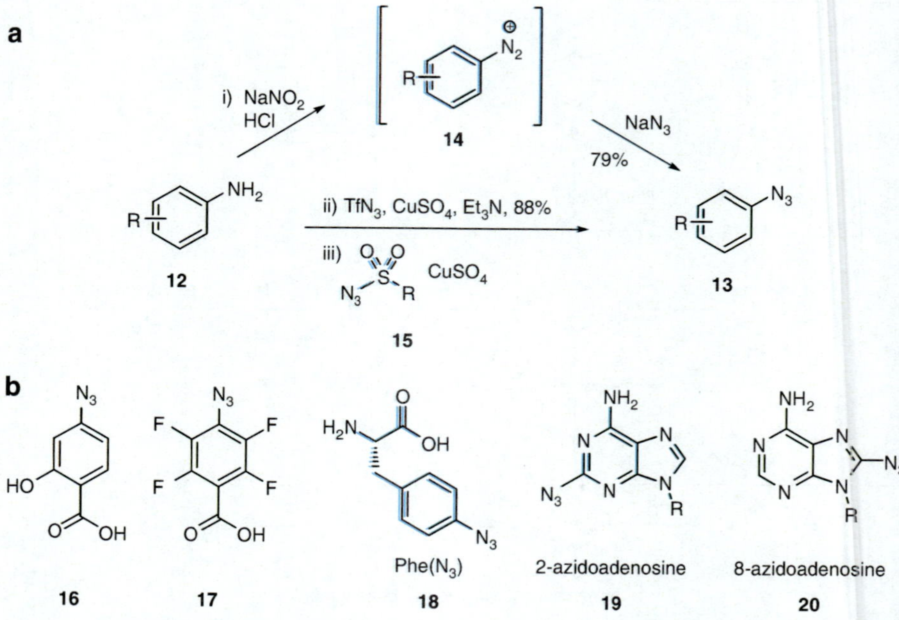

Scheme 2 Preparation and examples of aryl azides. (**a**) Three possible routes for the conversion of an aryl amine into its aryl azide: (*i*) via diazotization [44], (*ii*) by the use of triflyl azide [45], and (*iii*) with sulfonyl azides **15** (R = imidazole [46] or R = benzotriazole [55]). (**b**) Examples of substituted aryl azides

properties. In general, most substituents *ortho* to the azide are to be avoided because they can lead to undesired cyclizations after photolysis [30]. It has been found that introduction of electron-withdrawing substituents such as nitro, hydroxyl, and acyl groups (e.g., see compound **16** [48] in Scheme 2b) has the dual effect of increasing the molar absorptivity and red-shifting the maximum absorption wavelength, both of which positively influence the photoreactive properties [34, 49]. In addition, it has been found that (per)fluorinated aryl azides (such as **17**) [50] rearrange more slowly from the singlet nitrene species to the benzazirine and dehydroazepine (see Scheme 1, compounds **5** and **7**, respectively), which leads to more efficient insertion reactions [35]. The main advantage of aryl azides is their relatively small size and the possibility to incorporate them into natural biological compounds such as phenylalanine **18** [51], adenosines **19** [52], and **20** [53] without significantly altering the original structure.

Due to the many possible reaction pathways after irradiation (including capturing of the reactive intermediates by the solvent) crosslinking yields are often low (<30%). Arguably, the popularity of the aryl azide moiety in PAL studies is based on its relative ease of preparation and incorporation rather than on its photochemical properties.

An extensive study of matrix metalloproteinases (MMPs) with the use of an aryl azide modified A*f*BP was recently reported by Dive and coworkers [54]. MMPs are

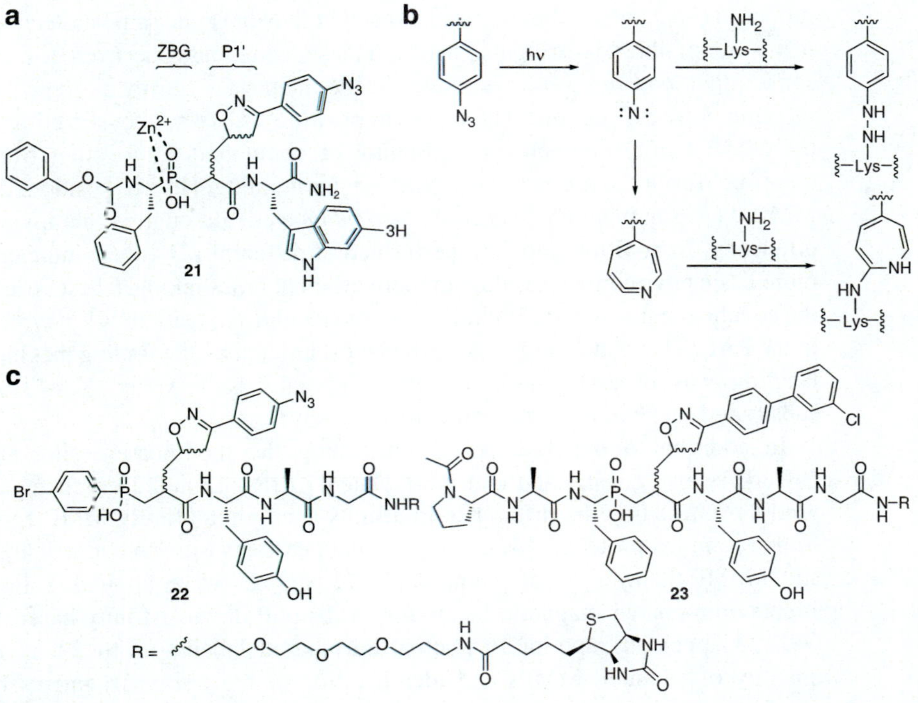

Scheme 3 Affinity-based probes targeting MMPs, from studies by Dive et al. [54]. (**a**) Photoreactive *Af*BP containing a tritium label. *ZBG* zinc binding group. (**b**) Possible constructs formed between hMMP-12 and **21** after photolysis proposed by Dive et al. [55]. (**c**) Structures of *Af*BPs with or without a photophore for pull-down of active MMPs

metalloproteases that reside in the extracellular matrix and are responsible for degradation of extracellular matrix material. Their mode of action depends on a Zn^{2+} ion in the active site, which coordinates the scissile bond carbonyl of the substrate and a water molecule. As a result, the carbonyl becomes more electrophilic and is subsequently hydrolyzed. The fact that there is no formation of a covalent bond between enzyme and substrate during the proteolysis makes this class of enzymes an interesting target for PAL. The authors describe the use of radiolabeled compound **21** (Scheme 3a), a potent, subnanomolar MMP inhibitor, to label and visualize purified human MMPs (hMMPs) [54]. The aryl azide photoreactive group is located at the P1′ pocket, which leads to a tight interaction with the enzyme's cavity. A large difference in terms of labeling efficiency and sensitivity between several MMPs was found, with MMP-12 giving the best results. The estimated crosslinking yield was ~42% after 2 min of irradiation, based on silver staining, and as little as 2.5 fmol MMP-12 could be detected. In a second study, the specifics of photocrosslinking were further explored, using compound **21** in combination with mass spectrometry and site-directed mutagenesis [55]. Interestingly, the ε-amine side chain substituent of Lys241 in MMP-12 appeared to play a crucial role in the photocrosslinking. Two possible covalent constructs were proposed (Scheme 3b), but due to the fact that they have the same molecular weight it was impossible to distinguish between these using mass spectrometry. In theory, some other constructs are possible (see Scheme 1), however, these were not mentioned by the authors. The lysine at position 241 is not conserved throughout the MMP family, and photocrosslinking of compound **21** to other MMPs was therefore further explored [56]. MMP-3 (containing His in position 241) and MMP-9 (Arg in position 241) could also be labeled, although with a lower overall efficiency. In addition, labeling performed at different pH values indicated that a more basic environment resulted in more efficient crosslinking. These results led to the conclusion that the nucleophilicity of the residue at position 241 plays a key role in the PAL. This conclusion was further substantiated by the finding that hardly any labeling was observed with hMMP-12 mutants K241A and K241T (Lys241 substituted by alanine or threonine, respectively).

In addition to the attempts at unraveling the modification site, Dive and coworkers also constructed two biotinylated A*f*BPs, **22** and **23** (Scheme 3c), and used these to study the difference in affinity- and photoaffinity MMP enrichment from a complex proteome [57]. For this, tumor extracts were spiked with hMMP-12 and hMMP-8, after which compounds **22** and **23** were applied, followed by streptavidin-coated magnetic beads for MMP pull-down. Affinity-based labeling with **23** appeared superior to photoaffinity-based labeling with **22** in terms of quantity of captured MMPs and identification of the tryptic fragments by mass spectrometry.

3.2 Diazirine

One of the major advantages of the photolabile diazirine group over aryl azides is that all its members absorb most efficiently at a wavelength of 350–380 nm. This is well above the 300 nm limit (vide supra) and therefore causes no significant damage to the biological system. The most important reactions that occur after photolysis of 3-aryl-3*H*-diazirines are shown in Scheme 4. When a diazirine (such as **24** or **25**) is irradiated, molecular nitrogen is expelled and a singlet carbene is formed (**26**). Competitively, a substantial amount (>30%) of the diazirine is converted into the diazoisomer **27**. This diazo compound can be converted into the singlet carbene under the influence of light; however, at the wavelengths normally used (360 nm) this process is relatively slow. For this reason the diazo species is relatively long-lived and thus has time to diffuse, resulting in either aspecific labeling or hydrolysis. This problem was largely tackled when Brunner and coworkers reported the development of 3-aryl-3-(trifluoromethyl)-3*H*-diazirine **25** [58]. The strong electron-withdrawing properties of the trifluoromethyl group stabilize the diazoisomer, which makes it almost completely resistant towards undesired "dark" reactions. Singlet carbene **26** is a very short-lived species ($t_{1/2}$ ~ 1 ns) and can be transformed into triplet carbene **34** via intersystem crossing. Singlet and triplet carbenes display a similar behavior compared to their corresponding nitrenes.

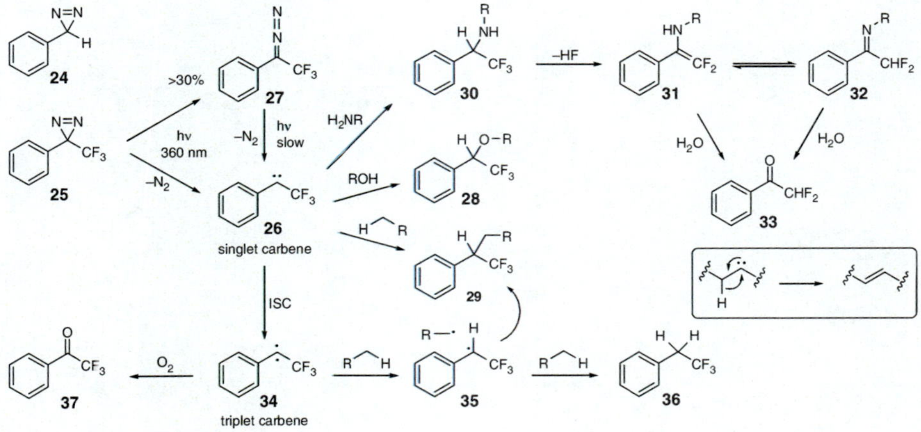

Scheme 4 Possible reactions of the intermediates formed after photolysis of 3-aryl-3*H*-diazirines. *Inset* shows olefin formation from carbene derivative of unsubstituted 3-alkyl-3*H*-diazirines via a hydride shift

A singlet carbene can react as an electrophile, nucleophile, or ambiphile, depending on the nature of its substituents, whereas triplet carbenes behave like diradicals. The singlet carbenes can give fast insertion reactions in which they hardly discriminate between different reaction sites. Insertions into hydroxyl groups (giving **28**) usually give a higher yield compared to C–H insertions (**29**) [59]. Insertion into a primary or secondary N–H bond (**30**) can lead to an undesired side reaction. The formed construct readily expels HF, thereby giving the enamine **31**, which is in equilibrium with imine **32**. In aqueous environments, such as a physiological sample, these species are subsequently hydrolyzed into the corresponding ketone **33**, with loss of the captured substrate as the result [59]. The triplet carbene can react with C–H bonds analogs to triplet nitrenes. Initial hydrogen abstraction leads to radical intermediate **35**, which either reacts with the formed carbon radical to give a net C–H insertion (**29**) or abstracts a second hydrogen from a different C–H bond, resulting in a reduction (**36**). Another undesired side reaction occurs when the triplet carbene is oxidized by molecular oxygen (a "notorious scavenger of triplet states") to the corresponding ketone **37** [34]. In general, unsubstituted 3-alkyl-3*H*-diazirines should be avoided since their corresponding carbenes are prone to hydride shift, which results in an olefin (see the insert in Scheme 4) [30].

Although the diazirine group itself is relatively small, aryl diazirines are quite bulky, but they can be incorporated into molecules with a structure similar to naturally occurring compounds. Furthermore, they are stable towards a wide variety of conditions, including strongly acidic, strongly basic, oxidating and reducing agents, which is a big advantage of diazirines compared to aryl azides. Drawbacks of diazirines are the formation of substantial amounts (>30%) of the diazo species after photolysis and the intrinsic efficient reactivity of the singlet carbene with O–H bonds, which often leads to scavenging of the reactive species by water. Also, the synthesis of diazirines is somewhat complicated compared to that of aryl azides.

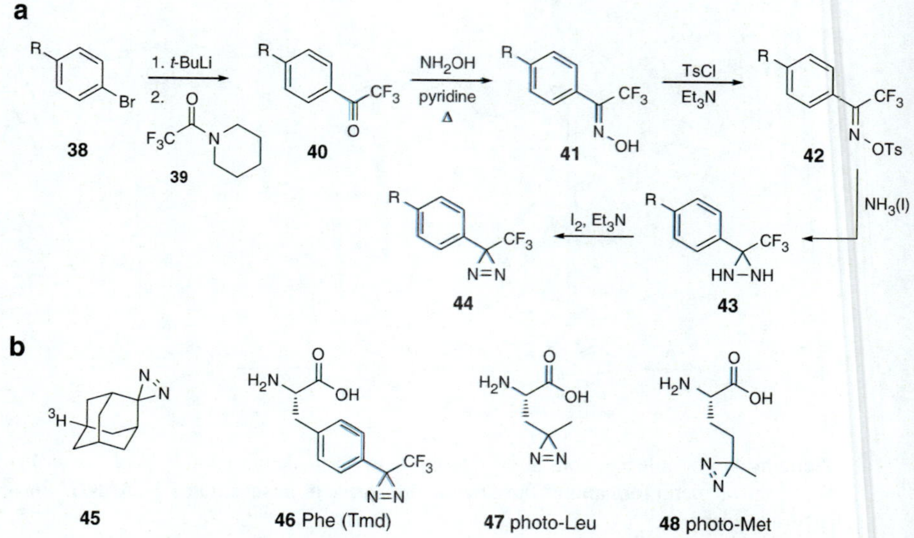

Scheme 5 Preparation and examples of diazirines. (**a**) Synthetic scheme for preparation of 3-aryl-3-(trifluoromethyl)-3*H*-diazirines. (**b**) Examples of diazirine functionalized compounds

The synthetic scheme often applied for the preparation of 3-aryl-3-(trifluoromethyl)-3*H*-diazirine nowadays is shown in Scheme 5a [60]. It starts by lithiation of an aryl bromide (**38**), which subsequently reacts with *N*-(trifluoroacetyl)piperidine **39** (easily prepared from trifluoroacetic anhydride and piperidine) under the formation of trifluoroacetophenone **40**. Next, the ketone is converted into the corresponding oxime **41**, after which the hydroxyl group is converted into its tosylate (**42**). Reaction with liquid ammonia (usually under pressure) allows the installment of the diaziridine group (**43**). Subsequent oxidation with iodine finally results in the diazirine (**44**). This five-step reaction sequence is especially compatible with acid-labile protective groups, which are often used to protect and/or install functionalities at the R position.

Some interesting examples of diazirines used in biologically relevant studies are shown in Scheme 5b. Tritium-functionalized adamantyl diazirine **45** was used for selective labeling of intrinsic membrane proteins in human erythrocytes. Despite the presence of α-hydrogen atoms, the formed carbene is not prone to hydride shift (see Scheme 4) due to the constraints of this caged ring system. However, photolabeling of this species is reported to be quite inefficient, probably due to its propensity to intramolecular C–H insertion reactions and reaction with water [30, 34]. Among the diazirine-functionalized amino acids developed, modified L-phenylalanine [Phe(Tmd), **46**] is the most popular. The first stereoselective preparation of this compound was reported by Nassal in 1984 [61] and it has been used extensively ever since [33, 36]. Recently, the synthesis of its D-phenylalanine analog was reported and this compound was used to probe the sweet taste receptor [62]. In 2005, Thiele and coworkers reported the chemo-enzymatic synthesis of diazirinized leucine (**47**) and methionine (**48**), which were

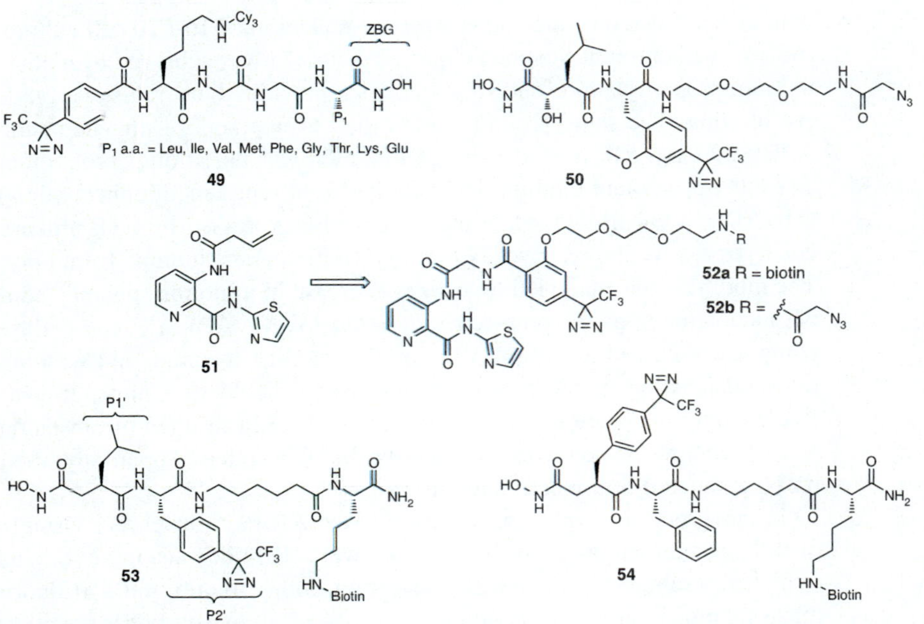

Fig. 4 Examples of A*f*BPs containing the diazirine moiety used to study active metalloenzymes

abbreviated as "photo-Leu" and "photo-Met". It was shown that these unnatural amino acids could be incorporated into proteins by a eukaryotic cell with genetically unmodified translational machinery, and this methodology was applied in the identification of protein–protein interactions in living cells [63]. The synthesis of photo-Met was optimized by Muir and coworkers in 2007, who circumvented the enzymatic resolution step and incorporated the unnatural amino acid into a protein using solid phase peptide synthesis and expressed protein ligation strategies [64].

Some interesting examples of A*f*BPs containing the diazirine moiety, which were used to target active metalloenzymes, are shown in Fig. 4. Yao and coworkers reported a library of hydroxamate oligopeptides **49**, with nine different amino acids at the P1 position [65]. The hydroxamate moiety is a strong zinc-binding group. The oligopeptides were modified with an N-terminal aryl diazirine for covalent modification of the target enzyme, and a fluorescent label (Cy3) for visualization. They were able to selectively label and visualize thermolysin (a Zn^{2+}-dependent metalloprotease found in Gram-positive bacteria) spiked in a crude yeast extract after covalent modification by irradiation for 20 min. In addition, the library of compounds was incorporated in a large-scale profiling study, in which the "fingerprint" labeling of 12 yeast metalloproteases towards probe library **49** was determined.

In two other studies, PAL of metalloenzymes was combined with two-step modification and visualization using the Cu(I)-catalyzed click reaction (see Fig. 2). Qiu et al. reported the use of succinylhydroxamate oligopeptide **50** containing an azide functionality in labeling MMP-2 (a secreted Zn^{2+}-dependent MMP) both as a purified enzyme and in a mouse melanoma B16-F10 cell culture medium [66]. Visualization of the photocaptured construct was achieved by a click reaction to biotinpropargylamide and subsequent streptavidin–horseradish peroxidase western blotting. The same group reported the development of one-step and two-step A*f*BPs **52a** and **52b** [67], the design of which was based on parent compound **51**. The latter is a potent inhibitor (IC_{50} 0.13 μM) of type I methionine aminopeptidase (MetAP1), a cobalt-dependent metalloenzyme expressed by both prokaryotic and eukaryotic cells that removes N-terminal (initiator) methionine from polypeptides. The modifications made led to a slight decrease in inhibitory potency compared to **51**. Incubation of overexpressed *Escherichia coli* MetAP1 in *E. coli* cell lysate with compound **52b** followed by UV irradiation, a click reaction, and western blotting revealed labeling of the target enzyme, which could be competed away by **51**. Interestingly, incubation with the one-step probe **52a** resulted in substantial nonselective labeling, which overwhelmed the MetAP1signal. Apparently, the two-step probe is much more selective in this case.

In our group, a number of diazirine-based A*f*BPs to label and visualize active metalloproteinases were developed as well. Peptide succinylhydroxamate **53** (which resembles compound **50**), featuring both a biotin and a trifluoromethyldiazirine moiety at the P2′ position was synthesized by solid phase peptide synthesis and checked for its ability to label several MMPs and their membrane-anchored relatives, ADAMs (a disintegrin and metalloproteinase) [68]. Upon incubation of purified recombinant ADAM-10 and subsequent irradiation with UV light (366 nm), the metalloprotease was covalently and irreversibly modified in an activity-based manner, as was evidenced by SDS-PAGE of the denatured protein followed by streptavidin blotting. The efficiency of the PAL for other MMPs and ADAMs, however, proved rather modest. This raised the question of whether the photoreactive group would be better directed towards the P1′ pocket, rather than the P2′ pocket (see Fig. 4). Examination of the available three-dimensional structures of metalloproteinase–inhibitor complexes indicated that the P1′ pocket in general is tightly binding and should be able to accommodate rather bulky hydrophobic groups at this position. We decided to address this issue by the synthesis of peptide hydroxamate **54** (Fig. 4) with the photoreactive group at P1′, and compare its MMP/ADAM labeling efficiency to that of probe **53** [69]. Indeed, compound **54** proved to be superior to **53** since it gave efficient labeling of ten recombinant MMPs and four ADAMs in an activity-dependent manner.

3.3 Benzophenone

A major advantage of benzophenones is that they can be excited at wavelengths of 350–360 nm, just like diazirines. The possible reaction pathways of benzophenones after photolysis are shown in Scheme 6. Absorption of a photon of the proper wavelength by a benzophenone (**55**) initially results in the formation of a triplet state benzhydril diradical (**56**). The formation of the triplet diradical is reversible

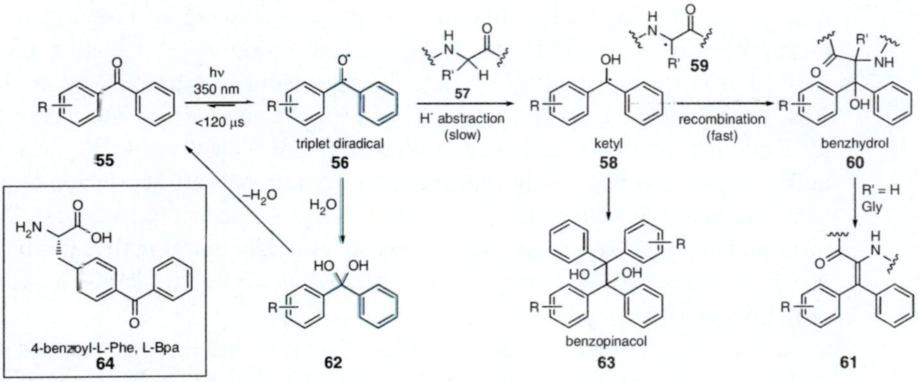

Scheme 6 Chemistry of benzophenones after photolysis

and it can exist as long as 120 μs before relaxing back to its ground state, in the case that it is unable to find a reaction partner. The first reaction step of the formed reactive species is hydrogen abstraction and the reaction rate is therefore dependent on the nature of a nearby X–H bond [70]. In general, the diradical is more reactive towards C–H bonds than towards O–H bonds. Those C–H bonds that form relatively stable carbon radicals are especially prone to react, and these include benzylic positions, amino acid α-positions, hydrogen atoms adjacent to heteroatoms, and tertiary carbon centers. Reactions with aromatic and vinylic C–H bonds have not been reported. All amino acids can react, although it has been shown that there is a preference for the ε-H in methionine when the benzophenone moiety is mobile enough to choose [71]. Abstraction of a hydrogen from an amino acid α-center (**57**) by **56** results in the formation of a ketyl (**58**) and an alkyl radical (**59**), which recombine fast to form a benzhydrol (**60**). In the case of glycine, there is a possibility of elimination of water under the formation of olefin **61**. A big advantage of the benzophenone group is that its photoactivated counterpart is more reactive towards C–H bonds compared to nitrenes, and is less prone to intramolecular rearrangements than carbenes. Also, when the diradical inserts into water the corresponding hydrate (**62**) is formed. This species quickly dehydrates to form the ketone (**55**) again, which can be recycled under irradiation to the diradical species. This ability of benzophenones to "search" for a good reaction center is a big advantage in terms of crosslink efficiency. However, when the reactive species is not quenched in time there is a big chance of nonspecific labeling, especially when a corresponding A*f*BP is not interacting with the target enzyme, but moves around freely in the medium.

A possible side reaction that can take place is the homodimerization of ketyl **58** to form benzopinacol **63**. However, due to the relatively big difference in reaction rates of hydrogen abstraction and recombination, normally only a very small amount of this is formed.

In contrast to aryl azides and diazirines, the most commonly used benzophenone building blocks are commercially available. Benzophenone-substituted amino acid analogs were also created, similar to those previously mentioned for aryl azides and diazirines. Not surprisingly, the most studied amino acid derivative is the one derived from phenylalanine, *p*-benzoylphenylalanine, commonly abbreviated as Bpa (**64** in Scheme 6) [72]. Although the benzophenone group may seem like the ideal photocrosslinking reagent, it also suffers from some drawbacks. It is relatively bulky, which can negatively influence the interaction between enzyme and substrate. Also, the resulting steric hindrance can give rise to discrimination between reaction sites and, as a result, lead to nonspecific labeling. Finally, irradiation for prolonged times (>30 min) is often needed in order to obtain a reasonable crosslinking efficiency [27].

In the recent literature, many examples of A*f*BPs containing benzophenones can be found. A first example concerns the study of HDACs. These enzymes catalyze the hydrolysis of acetylated lysine amine side chains in histones and are thus involved in the regulation of gene expression. There are approximately 20 human HDACs, which are divided into three classes (I, II, and III). Class I and II HDACs are zinc-dependent metallohydrolases that do not form a covalent bond with their substrates during their catalytic process, which is similar to MMPs. It has been found that hydroxamate **65** (SAHA, see Fig. 5) is a potent reversible inhibitor of class I and II HDACs. In 2007, Cravatt and coworkers reported the transformation of SAHA into an A*f*BP by installment of a benzophenone and an alkyne moiety, which resulted in SAHA-BPyne (**66**) [73]. They showed that the probe can be used for the covalent modification and enrichment of several class I and class II HDACs from complex proteomes in an activity-dependent manner. In addition, they identified several HDAC-associated proteins, possibly arising from the tight interaction with HDACs. Also, the probe was used to measure differences in HDAC content in human disease models. Later they reported the construction of a library of related probes and studied the differences in HDAC labeling [74]. Their most

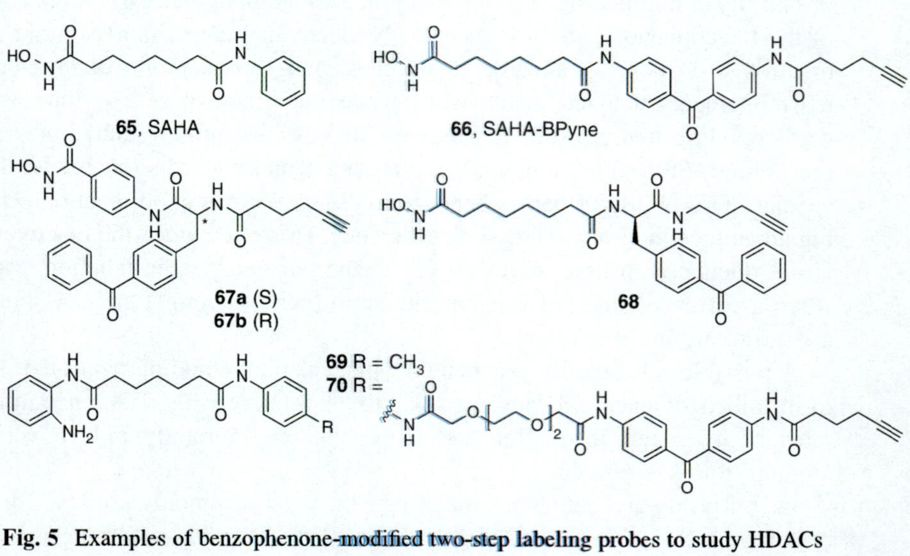

Fig. 5 Examples of benzophenone-modified two-step labeling probes to study HDACs

interesting finding was that the labeling efficiency is not directly linked to a compound's inhibitory potency. For example, compounds **67a**, **67b**, and **68**, containing Bpa (Fig. 5), showed a higher potency (in terms of inhibition of HDAC activities) than **66**; however, the latter compound proved to be superior in HDAC labeling. A similar approach was reported by Xu et al. in 2009, in which the potent HDAC class I (comprising HDACs 1, 2, 3, and 8) inhibitor **69** was modified with a benzophenone-spacer-alkyne moiety (**70**) to study its binding affinities in more detail [75]. Incubation of Friedreich's ataxia lymphoblast-derived nuclear extract with **70**, followed by photoaffinity crosslinking and a click reaction with either rhodamine azide or biotin azide, identified HDAC-3 as the single target of this inhibitor.

In addition to the aryl azide and diazirine groups, the benzophenone moiety was also incorporated in probes that target MMPs (see Fig. 6). Potent, broad-spectrum succinyl-hydroxamate MMP inhibitor GM6001 (**71**) was converted into photoaffinity probe **72** by incorporation of Bpa (**64**, Scheme 6) and a fluorophore, as reported by Cravatt and coworkers in 2004 [76]. This probe can be used to covalently label (through photocrosslinking) and visualize several active MMPs in complex proteomes. In addition, the authors were able to identify a number of other metalloproteases targeted by GM6001 that do not belong to the MMP family. In order to address this metalloenzyme's lack of selectivity towards a single inhibitor, this same group developed an alternative profiling strategy [77]. A two-step PAL MMP probe library was constructed in which the Bpa moiety was incorporated at either the P3′ (**73a,b**) or P2′ (**74**) position, and the P1′ and P2′ substituents were varied (see Fig. 6). The library compounds were applied to proteomes as a "cocktail" and instead of identifying the affinity of each

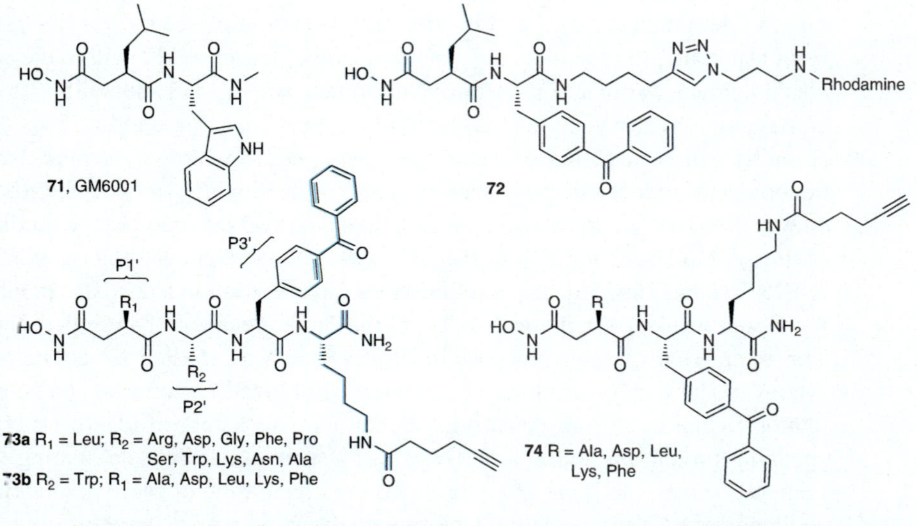

Fig. 6 **Examples of benzophenone-modified probes to target matrix metalloproteinases.** The amino acid three letter abbreviations in the R_1 (**73a,b**) and R (**74**) substituents refer to their corresponding side chains

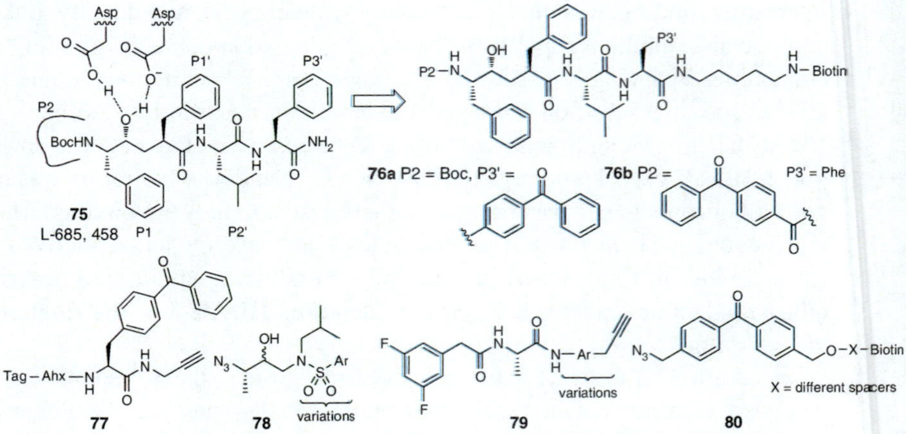

Fig. 7 Examples of A*f*BPs modified with the benzophenone photophore that target γ-secretase

metalloprotease towards a single probe the labeling profiles towards the entire library were analyzed collectively. This method proved to be powerful in that more than 20 metalloproteases (MMPs and others) could be identified from a complex biological mixture. In a later study, the authors applied some of the library compounds and related photoaffinity probes as competitive A*f*BPs to study the affinity of four MMP-13 inhibitors for a large number of other metalloproteases [78].

A powerful application of the use of benzophenone-containing A*f*BPs to locate the active site within a protease complex has been reported for γ-secretase [79]. This multisubunit, integral membrane protein complex is responsible for the proteolysis of transmembrane proteins. Together with β-secretase, it generates the amyloid β-protein, which is known as the central pathogenic feature in Alzheimer's disease. Although the nature of the γ-secretase catalytic activity was determined to be of aspartate protease type, the exact location of its active site within the complex was unknown. A potent inhibitor of γ-secretase activity is L-685,485 (**75**, Fig. 7), containing a hydroxyethylene dipeptide-isostere. This compound mimics the transition state of the aspartate protease catalytic process and, hence, forms a reversible, noncovalent adduct with the active enzyme (the binding of the inhibitor in the enzyme's active site is shown in Fig. 7). Li and coworkers reported the modification of the inhibitor's P2 and P3′ substituents with a benzophenone moiety, which led to A*f*BPs **76a** and **76b** [79]. The modifications did not result in a decrease in inhibitory potency towards γ-secretase activity inhibition compared to parent compound **75**. The interaction of both probes with γ-secretase was studied by addition of the probes to HeLa cell membranes containing solubilized γ-secretase and subsequent photocrosslinking. From the results the authors were able to identify membrane-spanning protein presenilin 1 (PS1) as the γ-secretase active site-bearing subunit. Four years later, the same group reported the preparation of the synthetically more challenging P1′ Bpa-containing biotinylated analog of **75** (structure not shown) [80]. With this probe, they were able to label active γ-secretase in living cells and their results demonstrated, for the first time, that active γ-secretase is presented on the cell surface.

The use of click chemistry has also been applied to the synthesis of benzophenone-modified γ-secretase probes. The group of Yao reported the preparation of a compound library built up from Bpa-containing alkyne **77** and azide **78** (Fig. 7) [81]. The azide part contains a racemic hydroxyethylene moiety, and variations were made in its aryl sulfonamide domain. The compound library was screened for its potency against γ-secretase inhibition and the most potent compounds were used to label active PS1 in a cell lysate. In addition, Fuwa and coworkers reported a divergent synthesis of γ-secretase A*f*BPs by means of click chemistry with alkyne **79** and azide **80** [82]. Variations were made in the aryl part of the alkyne (dibenzoazepine or benzodiazepine) and in the type of spacer between the benzophenone moiety and biotin in the azide. PAL using these probes provided the authors with evidence that the molecular target of this type of probe is the N-terminal fragment of PS1.

Another class of enzymes that is especially well suited for PAL is the kinase family. These enzymes catalyze the ATP-dependent phosphorylation of several substrates, but do not form a covalent linkage with either reaction partner and can therefore not be caught by means of a suicide trap. Some examples of photoreactive A*f*BPs targeting kinases are shown in Fig. 8. In 2003, the group of Sewald reported the preparation of fluorescently tagged A*f*BP **82** [83], which was obtained after modification of the potent kinase inhibitor H-9 (**81**), an isoquinolinesulfonamide-containing competitive inhibitor that targets a broad range of kinases by occupying their ATP binding site. The probe proved to be capable of labeling several kinases in a concentration-dependent manner and could be eliminated by preincubation of the kinase with competing ligands. Furthermore, the authors were able to label,

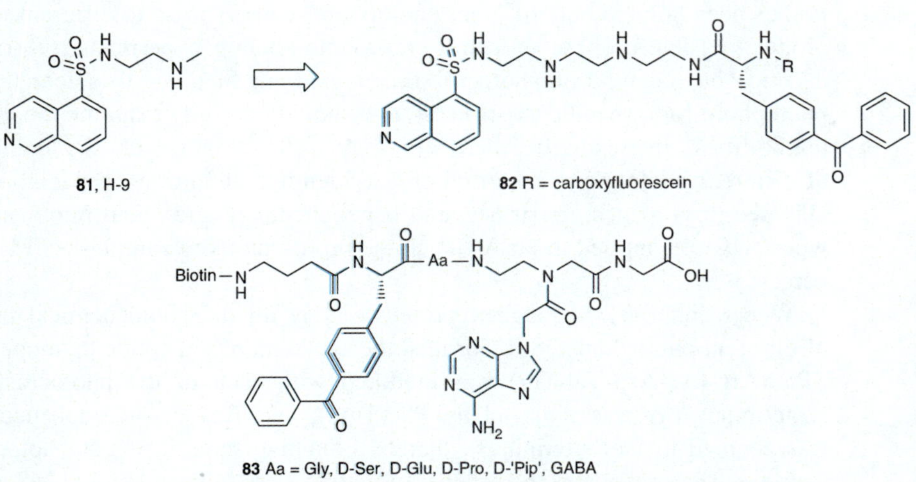

Fig. 8 Examples of benzophenone-containing A*f*BPs for labeling of kinases. D-"*pip*" (*R*)-piperidine-2-carboxylic acid, *GABA* γ-aminobutyric acid

although not very selectively, creatine kinase added to a mixture of isolated thylakoid proteins.

In 2006, Kawamura and coworkers reported a study on the photoactivated labeling of kinases with compound **83** (Fig. 8, with the amino acid being glycine) [84]. In an initial screen using six different kinases it was shown that this probe selectively labeled one kinase, namely leukocyte-specific protein tyrosine kinase (Lck) and that the labeling could be blocked by adenine. In addition, the probe selectively labeled Lck in an extract of Jurkat cells. With the aid of liquid chromatography coupled with tandem mass spectrometry (LC-MS/MS) after tryptic digest, it was possible to identify the labeled fragment within the kinase. Photocrosslinking had taken place in the Ile379–Arg386 tryptic fragment and, more precisely, to either Gly383 or Leu384. In a later study, the influence of target-binding affinity and conformational flexibility on the photocrosslinking efficiency was assessed [85]. For this, a small library was prepared in which the glycine moiety in **83** was replaced by other (non-proteinogenic) amino acids. Variations in length, flexibility, and hydrogen bonding capability were made. The results corroborated the earlier mentioned finding (see the examples of HDAC inhibitors) that the inhibitor potency does not necessarily correlate with the photocrosslinking efficiency. Interestingly, it was found that higher crosslinking yields were obtained for the more flexible compounds, whereas in other cases (for other enzymes) crosslinking efficiencies increased with a more tightly bound probe (see the examples of MMP probes in Sect. 3.2), however, this probably depends on the type of photocrosslinker used as well as the class of enzymes studied.

3.4 Comparing Photocrosslinkers

As has been pointed out, different photophores can give rise to different labeling products. Therefore, the outcome of a photolabeling experiment substantially depends on the type of photocrosslinker applied. In order to select the ideal photophore for a specific experiment, one should carefully examine the different possibilities. Interestingly, there are only a few reports on the comparison of different photophores under otherwise identical conditions, especially where all three (aryl azides, diazirines, and benzophenones) are taken into account or when they are applied in an A*f*BP. Some representative examples will be given here.

Weber and Beck-Sickinger reported a study on the photochemical behavior of different photophores [86]. For this, the synthetic pentapeptide thymopentin, or TP5 (Arg-Lys-Asp-Val-Thr) was modified with each of the photocrosslinkers (C-terminal Tyr was replaced by Phe(Tmd), and Bpa or 4-azidobenzoic acid was coupled to the N-terminus), thereby obtaining three different photoreactive peptides. The conversion and product formation were followed by LC-MS analysis after photolysis in water, *n*-propanol, and a water/*n*-propanol mixture (1:1, v/v). It was found that both the diazirine and aryl azide were quickly converted into a reactive species, but only the diazirine gave a relatively pure product resulting from insertion into the solvent, predominantly into the O–H bond. The aryl azide also gave a clean O–H insertion reaction in water. However, when *n*-propanol was present numerous unidentified products were formed, probably arising from intramolecular rearrangements and insertion reactions. The benzophenone-containing peptide gave clean C–H insertions in the presence of *n*-propanol, but the conversion yield was low after 15 min of irradiation (68% and 82% starting material left in water/*n*-propanol and *n*-propanol respectively). In pure water, the conversion was much higher (94% after 15 min); however, due to the poor reactivity towards O–H bonds, many unidentified products were generated. This led to the conclusion that the use of diazirines and aryl azides is preferred over benzophenones; however, the latter should be used in case the crosslinking site contains many water molecules.

Another direct comparison between the photophores, although not in an activity-based setting, was reported by Tate and coworkers [87]. They synthesized dUTP analogs containing four different photoreactive moieties **84a–d** (Fig. 9a) and incorporated them enzymatically into DNA constructs. With these, the DNA PAL of yeast RNA polymerase III transcription complexes was studied. It was shown that photolabeling with the diazirine construct (**84d**) rendered many protein–DNA contacts, whereas labeling with the other three photoreactive moieties (**84a–c**) proved only marginal.

In a recently reported study by Dalhoff and coworkers, the aryl azide and diazirine photophores were directly compared in an activity-based labeling of methyltransferases (MTases) [88]. This class of enzymes catalyzes the transfer of the activated methyl group in *S*-adenosyl-L-methionine (SAM) to a nucleophilic position in various substrates (DNA, RNA, proteins) under the formation of *S*-adenosyl-L-homocysteine (SAH), as shown in the insert in Fig. 9b. SAH itself is a general competitive product inhibitor of MTases and because of this it was transformed into four A*f*BPs by attachment of a spacer moiety containing biotin and either an aryl azide or aryl diazirine (**85a,b**, Fig. 9b). Since the optimal modification site was unknown, the linker moiety was attached to either the N6 or C8 position, as indicated in **85**. Initially, the compounds were used to label and capture (by means of streptavidin-coated magnetic beads) purified DNA adenine N6 MTase from *Thermus aquatis* (M.TaqI). All four compounds could be used for M.TaqI pull-down in a yield of 5–10%. However, the capture efficiency of the diazirine-based probes was slightly higher compared to the aryl azides, as determined by western blotting. Labeling proved to be activity-based since it could be competed out with SAH and, in addition, was shown to be light-dependent. With these results in hand, the authors also tested the N6-aryl azide-derived probe **85a** for pull-down of three other NTases in their purified form and two other SAH binding proteins in *E. coli* lysate, which were identified by LC-MS/MS analysis. Unfortunately, the diazirine-derived probes were not further examined because the authors argued that “the small improvement in capture yield might not balance the much higher cost of preparation”.

From these studies, it seems that the diazirine group is the most effective photocrosslinker. Interestingly, there are two reports showing that the potency of

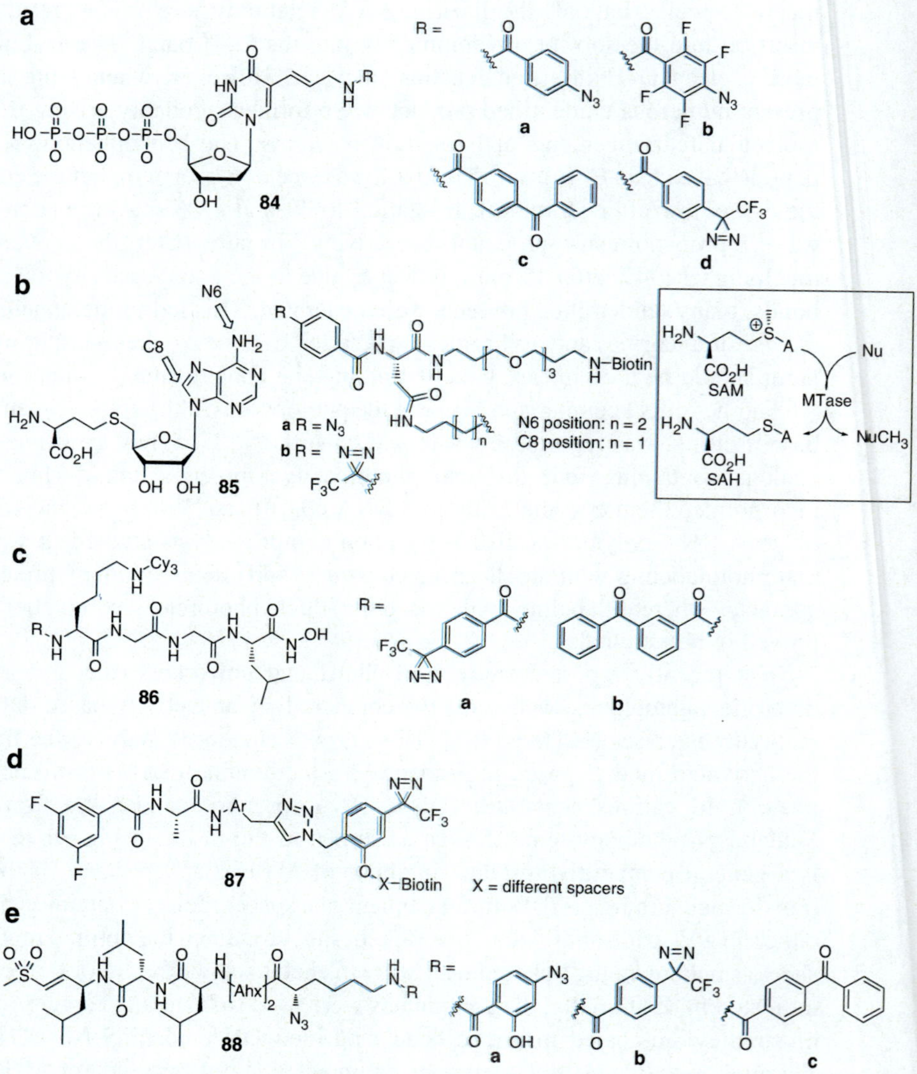

Fig. 9 Structures of compounds used to compare the photoaffinity labeling of different photoreactive groups. (**a**) Photoreactive dUTP analogs **84a–d** reported by Tate et al. [87]. (**b**) Photoreactive SAH analogs **85a,b** reported by Dalhoff et al. [88]. The *insert* shows the transfer of the activated methyl group from SAM to nucleophilic positions (Nu) in various substrates by methyltransferases (*MTase*). *A* adenosine. (**c**) Photoreactive metalloprotease probes **86a,b** reported by Yao et al. [65]. (**d**) Diazirine-containing analog of clicked compound **79–80** for labeling of γ-secretase, reported by Fuwa et al. [82]. (**e**) Photoreactive peptide vinyl sulfones **88a–c** used to probe the 20S proteasome cavity, reported by Geurink et al. [89]

the applied photoreactive group largely depends on the system it is used in. In a study of photoreactive metalloprotease probes by Yao et al. [65] (as outlined in Sect. 3.2) an additional benzophenone-containing probe was synthesized. In order to compare the diazirine and benzophenone moieties, the authors replaced the diazirine moiety in compound **86a** (Fig. 9c; this is the same compound as **49**, shown in Fig. 4, with P1 = Leu) with the corresponding benzophenone counterpart (**86b**). This compound proved to label purified thermolysin as well; however, no signs of labeling were observed when the compound was applied to crude yeast extract spiked with thermolysin. In this case, the diazirine-based probe appeared superior compared to its benzophenone counterpart. In contrast, Fuwa and coworkers obtained the opposite results, although in a completely different system [82]. In their study on the development of benzophenone-derived photoreactive γ-secretase probes (see Sect. 3.3 and Fig. 7), they also incorporated the diazirine photophore into their probes (**87**, Fig. 9d). These diazirine-containing analogs of the compounds resulting from a click reaction between **79** and **80** (Fig. 7), proved to be completely ineffective in the labeling of γ-secretase active subunits, despite the fact that they showed the same inhibitory potential compared to their benzophenone counterparts.

Recently, in our group we created a set of bioorthogonal photoreactive suicide substrates **88a–c** and used these to probe the proteolytically active 20S proteasome core particle [89]. The ABPs are equipped with a C-terminal vinyl sulfone moiety for the covalent and irreversible modification of active proteasome subunits. Successively, the ABPs can be crosslinked via a photophore (located at the N-terminus) to residues distal to the active site reactive group. The azide was used for visualization and enrichment of the bound constructs after Staudinger–Bertozzi ligation. This provides information on the orientation that the inhibitor adapts within the 20S cavity. We opted to compare different photophores and installed the three most commonly applied ones. All three ABPs showed equal potency towards the active proteasome subunits but, interestingly (and unexpectedly), of these three ABPs, the aryl azide-containing probe **88a** proved to be superior over the other two since it was the only one able to crosslink multiple subunits. LC-MS/MS analysis of the tryptic digest of the crosslinked complex revealed a linkage between the β5 and β6 subunits.

These examples show clearly that there is no such thing as the "ideal photophore". Due to the differences in reaction site preferences and the nature of the reactive intermediates of each individual moiety, the optimal photophore type needed differs from system to system. In most cases, it can be worthwhile to spend some time on the optimization of the used photophore in a certain A*f*BP in order to obtain the best results. In addition, interesting information can be obtained when the results derived from different photophores are combined [34].

3.5 *Identification of the Photocrosslinked Sites*

As outlined in Sect. 2, the tag incorporated in an ABP or A*f*BP is used for visualization, both in vivo and for read-out after SDS-PAGE, of the bound enzyme. In addition, a bifunctional tag such as biotin can not only be used as a visualization tool but can also be used as a pull-down handle for purification of the covalently modified complex. Subsequent (tryptic) digestion and LC-MS/MS analysis allows the identification of the tryptic peptides and, hence, the labeled enzyme(s). If a warhead-containing ABP is used, the modified peptide fragments can in general be found easily in the LC-MS/MS analysis, since the chemistry of modification and, hence, the nature of the final construct is known. In PAL, the identification of the photolabeled fragment is usually much more difficult. This is largely due to the fact that the site of modification is unknown (usually there are multiple modification sites) and the photocrosslinking reaction itself can lead to many possible products, because of the multiple reactive species involved. In some cases, the identification of the photolabeled site is possible [see for instance the labeling of hMMP-12 reported by Dive et al. [54] (Sect. 3.1) or the labeling of Lck reported by Kawamura and colleagues [84] (Sect. 3.3)], but this is generally difficult because of complicated mixtures. A possible solution to this problem is the mixed isotope PAL strategy, which makes use of "MS-friendly" photoprobes, or target identification probes (TIPs) [90]. The general idea of this strategy is schematically shown in Fig. 10. A proteome is treated with a 1:1 mixture of two structurally identical A*f*BPs, which only differ in their absolute mass by incorporation of stable isotopes (e.g., deuterium, ^{13}C, ^{18}O, or ^{15}N). Subsequent photocrosslinking, purification (using the tag) and tryptic digestion leads to a mixture of labeled and unlabeled tryptic peptides. It is now possible to discriminate between these two types of peptides by LC-MS/MS analysis because of the double peaks separated by the mass difference between the "light" and "heavy" isotopically labeled species. Thus, the modified peptide(s) can be identified by searching for the unique isotopic pattern, or "isotope signature". Although this concept is relatively new in the field of PAL, some examples have been recently reported.

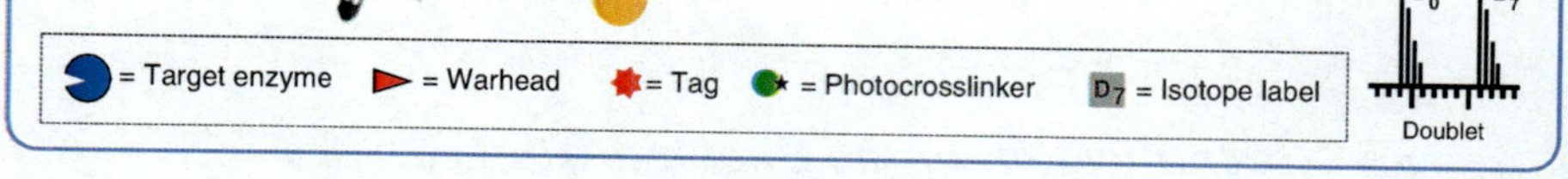

Fig. 10 Mixed isotope photoaffinity labeling strategy for determination of the labeling site. After binding and photocrosslinking the target enzyme with a 1:1 mixture of the "light" (D_0) and "heavy" (D_7) isotopically labeled A*f*BPs, the construct is purified and tryptically digested. LC-MS/MS analysis of the peptide pool allows discrimination between the labeled and unlabeled peptides. The modified fragment can easily be retrieved and identified by searching for the isotope signature

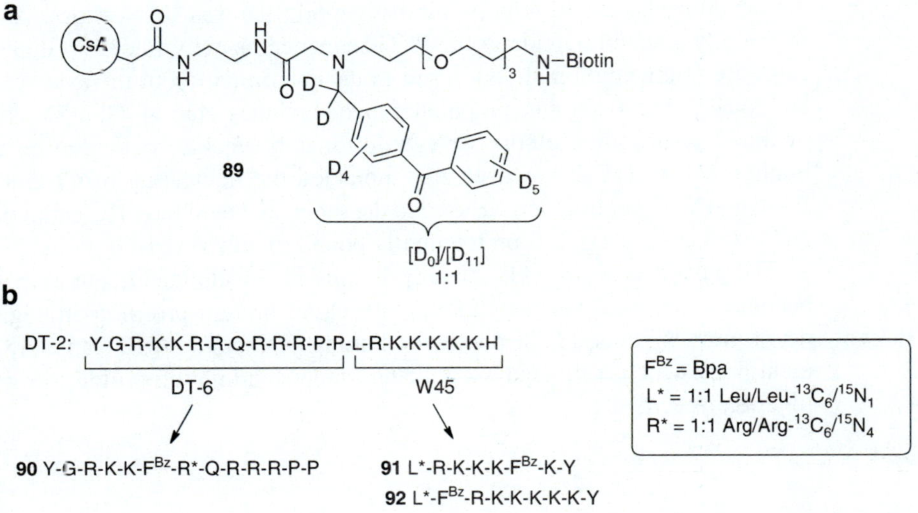

Fig. 11 Examples of 'MS-friendly' photoreactive compounds used in mixed photoaffinity labeling. *CsA* cyclosporin A

Lamos and colleagues reported the modification of cyclophilin A (CypA) by binding immunosuppressive cyclosporin A with a benzophenone-D_{11} and a biotin moiety (compound **89**, Fig. 11a) [91]. As a proof of principle, they used a 1:1 mixture of this TIP and its nondeuterated isoster for the selective PAL and pull-down of CypA among three other proteins. Subsequent tryptic digestion of the elutes and LC-MS/MS analysis allowed the identification of 11 CypA characteristic peptides, two of which were modified with the probe, as evidenced from the double, 11 Da separated, peaks in the mass spectra. The large 11-Da mass difference allowed easy visual recognition of labeled peptides in the mass spectra, which makes this a powerful method for determination of the modification site after PAL; however, application in more complex systems still remains to be done.

A second, more extensive, study was recently reported by the group of Heck [92]. These authors describe the application of peptides containing stable isotopes and a photoreactive moiety in the localization of cGMP-dependent protein kinase (PKG) substrate binding sites. For this, they focused on a potent and selective inhibitor of PKG, oligopeptide DT-2 (Fig. 11b), which is a construct of a PKG tight binding sequence (W45) and a membrane translocating sequence (DT-6). Both of these peptides were modified by incorporation of a benzophenone moiety (as Bpa) and an isotopically labeled amino acid Leu-$^{13}C_6/^{14}N_1$ or Arg-$^{13}C_6/^{15}N_4$ (peptides **90–92**; notice the difference in the Bpa position in the two W45-derived peptides). LC-MS/MS analysis was performed on photocrosslinked complexes of PKG with each of the peptides. The power of the strategy was reflected in the results, which led to the identification of each peptide's binding site. For example, the binding site of compound **92** was located in tryptic peptide Gln195–Arg203 and fragmentation of this sequence identified Met201 as the only crosslinked amino acid. Interestingly, the binding site of the highly similar peptide **91** was located on a completely differently situated peptide in the PKG sequence, namely Phe359–Glu374. In this case, the fragmentation did not result in the determination of the exact crosslinked amino acid, but it could be pinpointed to the residues Thr364–Glu374. With all the obtained results the authors were unfortunately unable to determine the exact binding site of DT-2; however, they obtained the interesting result that both the DT-6 and W45 peptides are targeted to the same pocket in the PKG catalytic domain and that therefore DT-2 is preferentially bound to dimeric PKG.

The mixed isotope PAL strategy is still in development, but is believed to become a powerful tool in photoaffinity-based protein profiling strategies in the near future. With its aid, the photocrosslinking site(s) of an A*f*BP can more easily be identified, which will lead to a better understanding of protein structure and function in general.

4 Outlook

The use of PAL in ABPP has been described here. The most often used photo-activatable groups were treated and attention was made to their chemistry post-photolysis. In addition, an overview was given of recently published examples of their application as part of an ABP or A*f*BP for the study of enzyme activities. The use of PAL proves to be especially powerful in cases in which the ABP and target enzyme do not form a covalent adduct. Although examples have been given in which the strength of both ABPP and A*f*BP were combined to obtain additional structural information on the target system, the main challenges in this field remain the improvement of the crosslinking efficiency and the identification of the modified moieties by mass spectrometry. Supported by the development of novel analytical technologies (such as the mixed isotope strategy), we believe that PAL will continue to make an impact on chemical biology research involving ABPP, in particular for the identification and study of enzymes for which no obvious alternative read-out systems in situ and in vivo are available.

References

1. Wasinger VC, Cordwell SJ, Cerpapoljak A, Yan JX, Gooley AA, Wilkins MR, Duncan MW, Harris R, Williams KL, Humpherysmith I (1995) Electrophoresis 16:1090
2. James P (1997) Q Rev Biophys 30:279
3. Cravatt BF, Wright AT, Kozarich JW (2008) Annu Rev Biochem 77:383
4. Evans MJ, Cravatt BF (2006) Chem Rev 106:3279
5. van Swieten PF, Leeuwenburgh MA, Kessler BM, Overkleeft HS (2005) Org Biomol Chem 3:20
6. Saxon E, Bertozzi CR (2000) Science 287:2007
7. Rostovtsev VV, Green LG, Fokin VV, Sharpless KB (2002) Angew Chem Int Ed Engl 41:2596

8. Speers AE, Adam GC, Cravatt BF (2003) J Am Chem Soc 125:4686
9. Baskin JM, Prescher JA, Laughlin ST, Agard NJ, Chang PV, Miller IA, Lo A, Codelli JA, Bertozzi CR (2007) Proc Natl Acad Sci USA 104:16793
10. Ning XH, Guo J, Wolfert MA, Boons GJ (2008) Angew Chem Int Ed Engl 47:2253
11. Poloukhtine AA, Mbua NE, Wolfert MA, Boons GJ, Popik VV (2009) J Am Chem Soc 131:15769
12. Willems LI, Verdoes M, Florea BI, van der Marel GA, Overkleeft HS (2010) Chembiochem 11:1769
13. Devaraj NK, Weissleder R, Hilderbrand SA (2008) Bioconjug Chem 19:2297
14. Rossin R, Verkerk PR, van den Bosch SM, Vulders RCM, Verel I, Lub J, Robillard MS (2010) Angew Chem Int Ed Engl 49:3375
15. Fonovic M, Verhelst SHL, Sorum MT, Bogyo M (2007) Mol Cell Proteomics 6:1761
16. Yasuda Y, Li ZQ, Greenbaum D, Bogyo M, Weber E, Bromme D (2004) J Biol Chem 279:36761
17. Bogyo M, Verhelst S, Bellingard-Dubouchaud V, Toba S, Greenbaum D (2000) Chem Biol 7:27
18. Greenbaum D, Baruch A, Hayrapetian L, Darula Z, Burlingame A, Medzihradszky KF, Bogyo M (2002) Mol Cell Proteomics 1:60
19. Liu YS, Patricelli MP, Cravatt BF (1999) Proc Natl Acad Sci USA 96:14694
20. Kidd D, Liu YS, Cravatt BF (2001) Biochemistry 40:4005
21. Adam GC, Sorensen EJ, Cravatt BF (2002) Mol Cell Proteomics 1:781
22. Tuin AW, Mol MAE, van den Berg RM, Fidder A, van der Marel GA, Overkleeft HS, Noort D (2009) Chem Res Toxicol 22:683
23. Lennon-Dumenil AM, Bakker AH, Maehr R, Fiebiger E, Overkleeft HS, Rosemblatt M, Ploegh HL, Lagaudriere-Gesbert C (2002) J Exp Med 196:529
24. Kessler BM, Tortorella D, Altun M, Kisselev AF, Fiebiger E, Hekking BG, Ploegh HL, Overkleeft HS (2001) Chem Biol 8:913
25. Verdoes M, Florea BI, van der Marel GA, Overkleeft HS (2009) Eur J Org Chem:3301
26. Kalesh KA, Shi HB, Ge JY, Yao SQ (2010) Org Biomol Chem 8:1749
27. Vodovozova EL (2007) Biochemistry (Moscow) 72:1
28. Tanaka Y, Bond MR, Kohler JJ (2008) Mol Biosyst 4:473
29. Singh A, Westheimer FH, Thornton ER (1962) J Biol Chem 237:3006
30. Bayley H (1983) Photogenerated reagents in biochemistry and molecular biology. Elsevier, Amsterdam
31. Chowdhry V, Westheimer FH (1979) Annu Rev Biochem 48:293
32. Schuster DI, Probst WC, Ehrlich GK, Singh G (1989) Photochem Photobiol 49:785
33. Brunner J (1993) Annu Rev Biochem 62:483
34. Fleming SA (1995) Tetrahedron 51:12479
35. Kotzybahibert F, Kapfer I, Goeldner M (1995) Angew Chem Int Ed Engl 34:1296
36. Dorman G (2001) Topics in current chemistry, vol 211. Springer, Berlin, 169
37. Fleet GWJ, Porter RR, Knowles JR (1969) Nature 224:511
38. Smith RAG, Knowles JR (1973) J Am Chem Soc 95:5072
39. Galardy RE, Craig LC, Printz MP (1973) Nat New Biol 242:127
40. Buchmueller KL, Hill BT, Platz MS, Weeks KM (2003) J Am Chem Soc 125:10850
41. Li YZ, Kirby JP, George MW, Poliakoff M, Schuster GB (1988) J Am Chem Soc 110:8092
42. Bergeron RJ, Dionis JB, Ingeno MJ (1987) J Org Chem 52:144
43. Staros JV, Bayley H, Standring DN, Knowles JR (1978) Biochem Biophys Res Commun 80:568
44. Shields CJ, Falvey DE, Schuster GB, Buchardt O, Nielsen PE (1988) J Org Chem 53:3501
45. Liu Q, Tor Y (2003) Org Lett 5:2571
46. Goddard-Borger ED, Stick RV (2007) Org Lett 9:3797
47. Katritzky AR, El Khatib M, Bol'shakov O, Khelashvili L, Steel PJ (2010) J Org Chem 75:6532
48. Aranda G, Beaucourt JP, Ponchant M, Isambert MF, Henry JP (1990) Eur J Med Chem 25:369

49. Geiger MW, Elliot MM, Karacostas VD, Moricone TJ, Salmon JB, Sideli VL, Stonge MA (1984) Photochem Photobiol 40:545
50. Pinney KG, Katzenellenbogen JA (1991) J Org Chem 56:3125
51. Wieland T, von Dungen A, Birr C (1971) Justus Liebigs Ann Chem 752:109
52. Czarnecki JJ (1984) Biochim Biophys Acta 800:41
53. Pomerantz AH, Rudolph SA, Haley BE, Greengard P (1975) Biochemistry 14:3858
54. David A, Steer D, Bregant S, Devel L, Makaritis A, Beau F, Yiotakis A, Dive V (2007) Angew Chem Int Ed Engl 46:3275
55. Dabert-Gay AS, Czarny B, Devel L, Beau F, Lajeunesse E, Bregant S, Thai R, Yiotakis A, Dive V (2008) J Biol Chem 283:31058
56. Dabert-Gay AS, Czarny B, Lajeunesse E, Thai R, Nagase H, Dive V (2009) Bioconjug Chem 20:367
57. Bregant S, Huillet C, Devel L, Dabert-Gay AS, Beau F, Thai R, Czarny B, Yiotakis A, Dive V (2009) J Proteome Res 8:2484
58. Brunner J, Senn H, Richards FM (1980) J Biol Chem 255:3313
59. Platz M, Admasu AS, Kwiatkowski S, Crocker PJ, Imai N, Watt DS (1991) Bioconjug Chem 2:337
60. Delfino JM, Schreiber SL, Richards FM (1993) J Am Chem Soc 115:3458
61. Nassal M (1984) J Am Chem Soc 106:7540
62. Masuda K, Koizumi A, Misaka T, Hatanaka Y, Abe K, Tanaka T, Ishiguro M, Hashimoto M (2010) Bioorg Med Chem Lett 20:1081
63. Suchanek M, Radzikowska A, Thiele C (2005) Nat Methods 2:261
64. Vila-Perello M, Pratt MR, Tulin F, Muir TW (2007) J Am Chem Soc 129:8068
65. Chan EWS, Chattopadhaya S, Panicker RC, Huang X, Yao SQ (2004) J Am Chem Soc 126:14435
66. Qiu WW, Xu J, Li X, Zhong L, Li JY, Li J, Nan FJ (2009) Chin J Chem 27:825
67. Qiu WW, Xu J, Li JY, Li J, Nana FJ (2007) Chembiochem 8:1351
68. Leeuwenburgh MA, Geurink PP, Klein T, Kauffman HF, van der Marel GA, Bischoff R, Overkleeft HS (2006) Org Lett 8:1705
69. Geurink PP, Klein T, Prely L, Paal K, Leeuwenburgh MA, van der Marel GA, Kauffman HF, Overkleeft HS, Bischoff R (2010) Eur J Org Chem:2100
70. Dorman G, Prestwich GD (1994) Biochemistry 33:5661
71. Wittelsberger A, Thomas BE, Mierke DF, Rosenblatt M (2006) FEBS Lett 580:1872
72. Kauer JC, Erickson-Viitanen S, Wolfe HR, Degrado WF (1986) J Biol Chem 261:695
73. Salisbury CM, Cravatt BF (2007) Proc Natl Acad Sci USA 104:1171
74. Salisbury CM, Cravatt BF (2008) J Am Chem Soc 130:2184
75. Xu CP, Soragni E, Chou CJ, Herman D, Plasterer HL, Rusche JR, Gottesfeld JM (2009) Chem Biol 16:980
76. Saghatelian A, Jessani N, Joseph A, Humphrey M, Cravatt BF (2004) Proc Natl Acad Sci USA 101:10000
77. Sieber SA, Niessen S, Hoover HS, Cravatt BF (2006) Nat Chem Biol 2:274
78. Nakai R, Salisbury CM, Rosen H, Cravatt BF (2009) Bioorg Med Chem 17:1101
79. Li YM, Xu M, Lai MT, Huang Q, Castro JL, DiMuzio-Mower J, Harrison T, Lellis C, Nadin JL, Neduvelil JG, Register RB, Sardana MK, Shearman MS, Smith AL, Shi XP, Yin KC, Shafer JA, Gardell SJ (2000) Nature 405:689
80. Chun J, Yin YI, Yang GL, Tarassishin L, Li YM (2004) J Org Chem 69:7344
81. Shi HB, Liu K, Xua A, Yao SQ (2009) Chem Commun:5030
82. Fuwa H, Takahashi Y, Konno Y, Watanabe N, Miyashita H, Sasaki M, Natsugari H, Kan T, Fukuyama T, Tomita T, Iwatsubo T (2007) ACS Chem Biol 2:408
83. Hagenstein MC, Mussgnug JH, Lotte K, Plessow R, Brockhinke A, Kruse O, Sewald N (2003) Angew Chem Int Ed Engl 42:5635
84. Hindi S, Deng HT, James L, Kawamura A (2006) Bioorg Med Chem Lett 16:5625
85. Kawamura A, Hindi S, Mihai DM, James L, Aminova O (2008) Bioorg Med Chem 16:8824

86. Weber PJA, Beck-Sickinger AG (1997) J Pept Res 49:375
87. Tate JJ, Persinger J, Bartholomew B (1998) Nucleic Acids Res 26:1421
88. Dalhoff C, Huben M, Lenz T, Poot P, Nordhoff E, Koster H, Weinhold E (2010) Chembiochem 11:256
89. Geurink PP, Florea BI, Van der Marel GA, Kessler BM, Overkleeft HS (2010) Chem Commun 46:9052
90. Robinette D, Neamati N, Tomer KB, Borchers CH (2006) Expert Rev Proteomics 3:399
91. Lamos SM, Krusemark CJ, McGee CJ, Scalf M, Smith LM, Belshaw PJ (2006) Angew Chem Int Ed Engl 45:4329
92. Pinkse MWH, Rijkers DTS, Dostmann WR, Heck AJR (2009) J Biol Chem 284:16354

Top Curr Chem (2012) 324: 115–136
DOI: 10.1007/128_2011_299
© Springer-Verlag Berlin Heidelberg 2011
Published online: 19 November 2011

Application of Activity-Based Protein Profiling to the Study of Microbial Pathogenesis

William P. Heal and Edward W. Tate

Abstract Activity-based protein profiling (ABPP) is a powerful technology for the dissection of dynamic and complex enzyme interactions. The mechanisms involved in microbial pathogenesis are an example of just such a system, with a plethora of highly regulated enzymatic interactions between the infecting organism and its host. In this review we will discuss some of the cutting-edge applications of ABPP to the study of bacterial and parasitic pathogenesis and virulence, with an emphasis on *Clostridium difficile*, methicillin-resistant *Staphylococcus aureus*, quorum sensing, and malaria.

Keywords Activity-based probes · Activity-based protein profiling · Antibiotic resistance · Catalomics · *Clostridium difficile* · Host–pathogen interactions · Infectious disease · Malaria · MRSA · Pathogens · Virulence factors

Contents

W.P. Heal and E.W. Tate (✉)
Department of Chemistry, Imperial College London, South Kensington Campus, London SW7 2AZ, UK
e-mail: e.tate@imperial.ac.uk

1 Introduction

1.1 Pathogenesis

During the later years of the nineteenth century, a scientific revolution occurred that gave us the field of microbiology. The ensuing advances in medicine and the plant and animal sciences, which saved countless lives and transformed agriculture, have rarely been matched in scale to this day. The more recent and very much on-going revolution started in the 1970s with the emergence of molecular genetics and accelerated with the sequencing of the genomes of many microorganisms. Now, in the post-genomic era, the functions of many microbial genes are still to be elucidated and many of the important questions concerning these organisms have yet to be answered.

Since the late 1990s, the development of proteomics, the study of protein structure and function, has played a vital role in turning our growing wealth of genetic data into a detailed picture of the processes within microorganisms and their interactions with their host. Indeed, the study of microbial pathogenesis has emerged as a vibrant scientific discipline in its own right, at a time when our ability to continue to combat infectious microorganisms is in some doubt. Steady evolution of antibiotic resistance is a perpetual challenge, with the discovery of novel classes of small molecule inhibitors struggling to keep up the pace.

1.2 Targeting Resistance

Microbial resistance has been documented against virtually every known therapeutic agent. Combined with our failure to discover new antibiotics at a competing rate, we are faced with the terrifying possibility of pre-penicillin levels of morbidity and mortality. Currently, 25,000 deaths annually in the European Union alone are due to antibiotic-resistant bacterial infection, mostly through iatrogenic infection [1]. This is against a background of a complete lack during the past few years of novel antibiotics that exploit new modes of action and a relatively low level of investment by the pharmaceutical industry due to the limited market value. If the current rise in resistance continues, however, this situation is set to undergo a radical reversal.

Understanding the mechanisms by which organisms develop resistance, often in diverse and surprising ways, is a vital part of the fight against these diseases, and can lengthen the lifespan of existing antibiotics. As discussed later, the β-lactam scaffold central to many key antibiotics and the target of widespread resistance has been used as an activity-based probe (ABP) in this context [2]. Potentially any antibiotic molecule could be transformed into an activity-based or affinity-based probe, allowing isolation and characterisation of enzymes that detoxify the drug.

As we will see, the probe design end of ABP technology [3, 4] can also provide us with entirely novel antimicrobial agents.

1.3 Targeting Virulence

Targeting virulence, the ability of the organism to establish infection and cause disease as opposed to its viability, is a relatively new paradigm in antimicrobial research [5]. Involving tightly controlled processes, including many allosterically activated signalling cascades, the production of virulence factors seems an insurmountably complex system to break down for study. For example, the phenomenon of quorum sensing and the detection of sufficient colony size to trigger different phases of growth and the release of host immune response suppressors is a field of intense activity [6]. In targeting virulence, the goal is to "disarm" the organism, stopping its ability to establish colonies, produce toxins, lyse red blood cells (haemolysins), piliate, upregulate certain proteases, form biofilms, or engage in other pathogenic behaviour, in the hope that evolution of resistance would be considerably slower than when targeting viability directly.

2 Profiling Enzymes Involved in Pathogenesis

2.1 Clostridium difficile *S-Layer*

Clostridium difficile is a Gram-positive antibiotic-resistant organism that flourishes in the intestinal region when competing bacteria that comprise the gut flora are eliminated by antibiotic treatment; asymptomatic colonisation is very common. There have been numerous high profile outbreaks around the world and many resultant deaths, with *C. difficile* being branded the quintessential "hospital superbug". Biochemical study of *C. difficile* has been hampered by a lack of genetic tools available for this organism.

The *C. difficile* cell is entirely covered by a paracrystalline protein surface layer [7], or "S-layer", that facilitates host–pathogen interactions and contributes to the organism's robustness. The lattice-like S-layer is comprised of high and low molecular weight S-layer protein chains (HMW and LMW SLPs, respectively) formed by cleavage of the full-length precursor protein SlpA. During growth, it has been estimated [8] that up to 400 molecules of SlpA are synthesised per cell per second, and translocated to the surface, cleaved and organised into the S-layer. The cleavage site is well conserved across many clinical *C. difficile* strains, suggesting that a similarly well-conserved protease is responsible. This putative protease is therefore extremely significant as a key feature of the pathogenesis of *C. difficile* and as a potential novel therapeutic target. However, until very recently, inhibitors

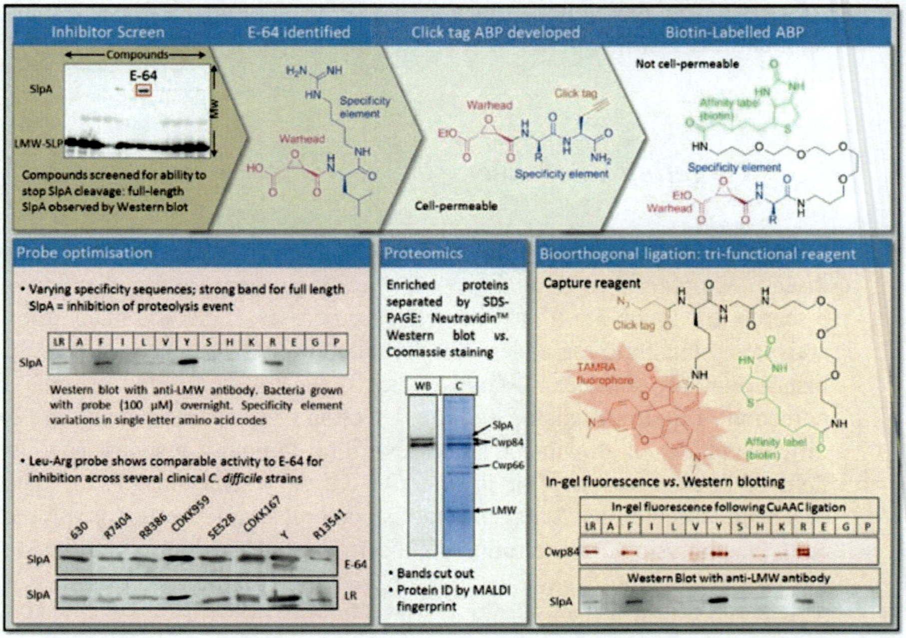

Fig. 1 ABPP in *C. difficile*

of S-layer processing were unknown, as was the identity and exact subcellular location of the enzyme responsible.

A recent study examined a range of commercially available protease inhibitors for their ability to inhibit S-layer processing [9]. The screen made use of a gel-based assay that uses an anti-LMW antibody to show the presence of both LMW and full-length SLPs. The authors qualitatively assessed the amount of unprocessed SlpA in glycine extracts of *C. difficile* 630 after incubation with the panel of inhibitors, discovering that E64, an epoxide-containing natural product broad-spectrum papain cysteine protease inhibitor [10], showed modest SlpA cleavage inhibition at 100 μM. Epoxide cysteine protease inhibitors usually attach covalently to the active site Cys residue proximal to the P_1 position in the substrate. With the E64 epoxide as the warhead, the structurally analogous Leu–Arg dipeptide-containing compound (Fig. 1) was prepared and found to provide equivalent inhibitory activity compared to E64. This allowed convenient solid-phase peptide synthesis technology to be applied to the exploration of the dipeptide structure–activity relationship (SAR). However, it was found that only the first residue was vital for inhibition, and thus only the preferred Leu was left upon attachment of propargyl glycine to form the first generation of "clickable" ABPs. These probes retain the activity observed in the natural product E64, but are small enough to enter the bacterium and label wherever the protease of interest might reside.

The use of the alkyne-containing E64-based ABP (Fig. 1), followed by bioorthogonal ligation to a novel doubly labelled reagent containing a TAMRA fluorophore and biotin, allowed labelling of an enzyme whose activity correlated with S-layer processing, along with a high level of background labelling. Although the Tyr-containing ABP, for example, still showed considerable SlpA cleavage inhibition at 10 μM, an order of magnitude enhancement over E64 inhibition, full-length SlpA was observed accumulating in the S-layer and only levelled off at an inhibitor concentration of 500 μM. This intriguing phenomenon was clarified upon inspection of the culture supernatants; above 250 μM inhibitor, SlpA is shed from the bacterium, reaching levels exceeding the total amount of S-layer attached to the cell wall. This implies that the tightly packed S-layer matrix does not abide full-length SlpA, and inhibition of cleavage leads to upregulation of SlpA biosynthesis. Indeed, it would appear that chemical intervention in SlpA processing has a dramatic effect on cell wall integrity. Due to the presence of the S-layer, healthy *C. difficile* is resistant to lysis by boiling in a strong anionic detergent. However, inhibitor-treated bacteria are highly susceptible to lysis.

As the target protein appeared to reside in the S-layer itself (the presence of SlpA in the S-layer implies that cleavage follows translocation), focus was moved away from the dual-stage approach (covalent alkyne tag attachment via epoxide warhead followed by subsequent click chemistry to add labels) to a single-step labelling method using a biotinylated, and apparently membrane-impermeable, E64 analogue ABP (Fig. 1). This allowed highly specific labelling which, with subsequent isolation and proteomic analysis, resulted in the identification of an entity that was putative cysteine protease Cwp84.

Utilising an S-layer processing mechanism reconstituted in *Escherichia coli*, it was possible to show that Cwp84 acts "on target", with specificity being largely determined by the P_2 Cys residue – most likely also the point of probe attachment. As well as de novo identification of Cwp84 via its catalytic activity and demonstration of its habitation of the surface layer, the ABPs were used to show conservation of the enzyme's activity across several strains of *C. difficile*. The work also produced the first inhibitors of S-layer biogenesis, providing an entry point for therapeutic development against this highly problematic organism. After considerable effort, a Cwp84 knockout *C. difficile* strain has also been established, proving conclusively that Cwp84 is the sole agent responsible for SlpA cleavage into the S-layer building blocks.

2.2 *Showdomycin-Based Probes*

Unlike the β-lactones and β-lactams, the mode of action of the unusual C-glucosyl nucleoside-based natural product showdomycin is unknown. Nevertheless, this compound has been shown to possess potent antibiotic properties, with results obtained in vitro suggesting a role as a suicide inhibitor of uridine metabolism [11]. Isolated from the bacteria *Streptomyces showdoensis*, showdomycin contains an electrophilic moiety, malaimide, in place of the base (cf. the structures of uridine or pseudouridine).

Böttcher and Sieber performed an adapted total synthesis of showdomycin that allowed the incorporation of 5-hexynoic acid as a clickable tag [11]. Using this tool, with identical antibiotic activity to showdomycin, they were able to profile the targets of showdomycin in *Staphylococcus aureus*, and compare enzyme profiles across different pathogenic strains, revealing strain-specific differences in regulation. Interestingly, maleimide is a potent electrophile, known to react indiscriminately with free thiols in Cys resides regardless of significance to catalytic mechanism. However, it would appear that the ribose creates a sterically hindered environment, limiting the off-target interactions that would otherwise be prevalent [12]. The authors confirmed this selective reactivity by incubating *S. aureus* lysate with both showdomycin and a simple fluorescent maleimide. The simpler probe labelled a multitude of targets, whereas the showdomycin probe was incorporated into only a very few targets. This, in addition to the good cell permeability, as demonstrated by feeding the probe to live pathogens and subsequent lysis, click chemistry, in-gel fluorescence imaging and gel-band mass spectrometry, suggest that the probe is a useful tool for profiling the targets of showdomycin (Fig. 2). Indeed, the probe was shown to be competitive for showdomycin binding sites (and not simply maleimide active), as pretreatment of proteomes with the antibiotic completely halted subsequent labelling attempts. Cell penetration and labelling of targets was found to be extraordinarily fast, requiring less than 10 min

In total, 13 different enzymes were identified from the four pathogens examined; namely *S. aureus*, methicillin-resistant *S. aureus* (MRSA), *Pseudomonas aeruginosa*

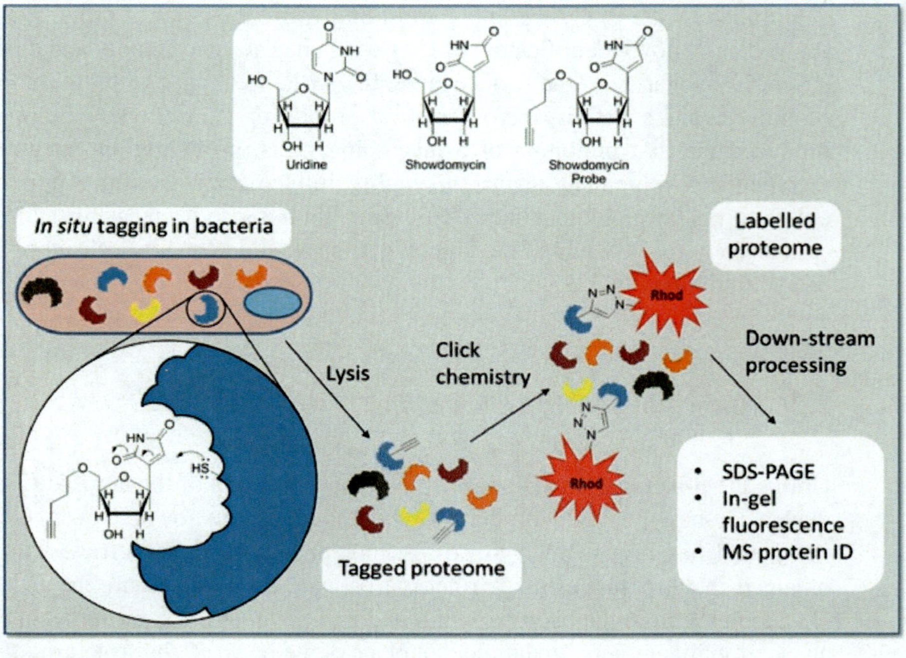

Fig. 2 Showdomycin probe design and labelling strategy

Table 1

Enzyme	Organism	Function
Phosphomethyl pyrimidine kinase (ThiD)	*S. aureus* NCTC 8325	Primary metabolism
Formate acetyltransferase (FAT)	*S. aureus* NCTC 8325, Mu50	
Cysteine desulfurase (CDS)	*P. aeruginosa*	
Alkyl hydroperoxide reductase subunit C (AhpC)	*S. aureus* NCTC 8325, Mu50, *P. Putida*, *P. aeruginosa*	Oxidative stress resistance
Alkyl hydroperoxide reductase subunit F (AhpF)	*S. aureus* Mu50, *L. monocytogenes* EGD-e	
Thioredoxine reductase (TrxB)	*S. aureus* NCTC 8325, Mu50	
Phosphoenolpyruvate protein phosphotransferase (PtsI)	*S. aureus* NCTC 8325, Mu50	Virulence
Acetolactate synthase (AlsS)	*S. aureus* NCTC 8325	
Inosine-5′-monophosphate dehydrogenase (IMPDH)	*S. aureus* NCTC 8325, Mu50, *L. monocytogenes* EGD-e	Nucleotide biosynthesis
Ribonucleotide diphosphate reductase subunit alpha (RNR)	*L. monocytogenes* EGD-e, *P. aeruginosa*	
UDP-*N*-acetylglucosamine 1-carboxyvinyltransferase 1 (MurA1)	*S. aureus* NCTC 8325, Mu50	Resistance and/or cell-wall biosynthesis
UDP-*N*-acetylglucosamine 1-carboxyvinyltransferase 2 (MurA2)	*S. aureus* Mu50	
UDP-*N*-acetylglucosamine pyrophosphorylase, putative (GlmU)	*S. aureus* NCTC 8325	

and *Listeria monocytogenes*. As shown in Table 1, the enzymes cover functions such as primary metabolism, oxidative stress resistance, virulence, nucleotide biosynthesis, and antibiotic resistance/cell wall biosynthesis. Major hits were proven to be labelled by the probe via recombinant expression and in vitro treatment.

The targets identified belong to families of oxidoreductases and transferases, many of which contain active-site Cys nucleophiles and react with nucleosides. Further confirmation of these enzymes' reactivity towards showdomycin was achieved in the case of MurA1 by an inhibition assay. This showed that showdomycin inhibited MurA1 with an IC_{50} (concentration resulting in 50% inhibition of activity) of 10 μM. In the case of AhpC, analysis of the site of modification revealed showdomycin to be attached only to Cys residues 39 and 168.

From the point of view of identification of enzymes involved in pathogenesis, the fact that this probe acts in a highly specific manner and fishes out targets across mechanistically distinct families shows just how useful this technique can be. The four different strains of *S. aureus* investigated with the showdomycin probe showed different enzyme profiles related to their pathogenic nature. Comparative analysis of MRSA strains Mu50 and DSM18827, highly toxin-producing clinical isolate DSM 19041, and reference strain NCTC 8325 showed that PtsI was labelled in both multiresistant strains and in the toxin-producing strain, but not in NCTC 8325. Conversely, peroxidase AhpC was found to be abundant in the reference strain, but not in the other three. Perhaps most interestingly, from the point of view of understanding the development of resistance of MRSA, labelling of MurA2 was only observed in multiresistant strain Mu50 and only in situ, not in vivo. The two copies of MurA, presumably the result of a gene duplication mutation, are functionally identical and one can replace the function of the other in the event of mutagenic interference. However, the organism cannot tolerate ablation of both genes. Despite the fact that all *S. aureus* species studied contain both genes (at comparable expressional levels, as determined by reverse transcriptase polymerase chain reaction), active MurA2 was only identified by the showdomycin probe in MRSA strain Mu50. This shows that despite conservation of genetic information across strains, post-translational regulation is responsible for observed variances in enzyme levels and thus for altered virulence and/or resistance. The work shows nicely how activity-based protein profiling (ABPP) can be used to detect key differences in proteomes that are otherwise indistinguishable.

2.3 Virulence Factors

2.3.1 Toxins of *C. difficile*

The principal virulence factors in *C. difficile* are the large glucosylating toxins TcdA and TcdB, and are thus of considerable interest as targets, especially TcdB, which is crucial for virulence and is found in all clinical isolates [13, 14]. Shen and coworkers recently reported the rational design of small molecule inhibitors of TcdB cysteine protease domain (TcdB CPD), and showed that ABPs based on these inhibitors could be used as tools for the study of virulence mechanisms [15]. Activation of the *C. difficile* CPD and subsequent glucosylating toxicity appear to be allosterically activated by sensing the eukaryotic cell environment. This is a complex and dynamic system of post-translational regulation and, as such, very challenging to study by traditional biochemical techniques.

The use of substrate-based peptide libraries coupled to either aza-epoxide or acyloxymethyl ketone (AOMK) electrophiles allows rapid access to an area of chemical space occupied by natural substrates. Looking initially at inhibitors known to be specific for a similar CPD in *Vibrio cholera* MARTX toxin ($MARTX_{Vc}$) [16], which have a tripeptide sequence attached to the warhead, the authors used a gel-based autocleavage assay (using a truncated TcdB(1–804) including CPDs and the natural autoprocessing site) to determine an SAR against TcdB CPD. However, weak inhibition was observed from the epoxides (e.g., Cbz-GluAla(aza)Leu-EP; Fig. 3), with only the AOMK containing compounds appearing to possess the preferred electrophilic reactivity. The SAR of carboxybenzyl-capped (Cbz-) inhibitors revealed that smaller or basic residues (such as Ala and Ser or Lys) were preferred by the P2 domain, with Cbz-AlaLeu-AOMK, Cbz-SerLeu-AOMK and Cbz-LysLeu-AOMK showing IC_{50}s between 1 and 2 μM. Further improvements in activity were achieved by changing the Cbz group to

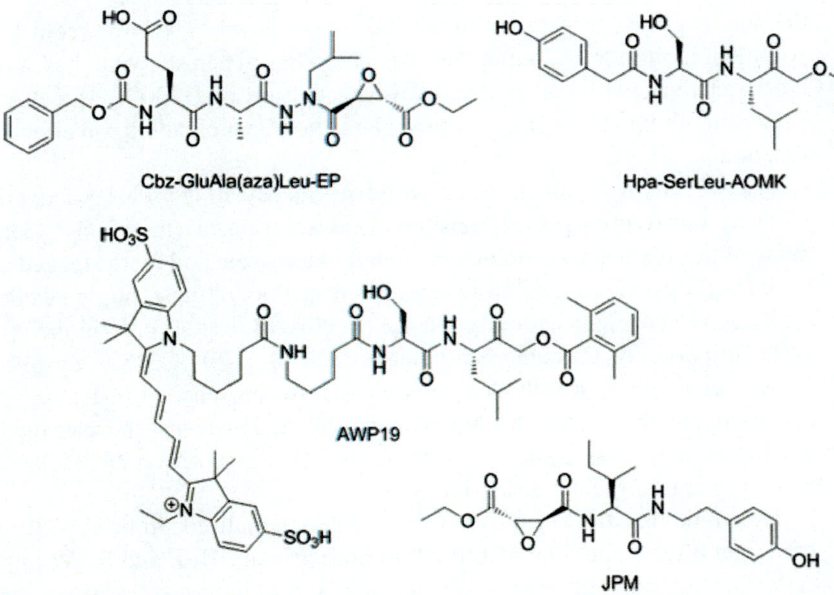

Fig. 3 Representative structures of epoxide inhibitors (Cbz-GluAla(aza)Leu-EP), AOMK inhibitors (Hpa-SerLeu-AOMK), fluorescent probe (AWP19, containing Cy5) and broad-spectrum cathepsin inhibitor JPM-OEt

hydroxyphenyl acetyl (Hpa-), with the inhibitor Hpa-SerLeu-AOMK (which contains amino acids found in the natural TcdB CPD substrate; Fig. 3) showing an IC_{50} of 0.71 μM. Further iterative optimisation of the P3 and P4 positions revealed that here small and or basic residues were also preferred. However, no compound outperformed the dipeptide inhibitor Hpa-SerLeu-AOMK.

For X-ray crystallography, the slightly less potent, but more soluble inhibitor Ac-GlySerLeu-AOMK was used, resulting in a 2 Å structure. This revealed that the P1 Leu is inserted far into the deep hydrophobic S1 pocket, appearing to make numerous van der Waals contacts, with the remaining inhibitor residues making minimal interactions. This hints at a common mechanism across bacterial CPDs, in which recognition of the P1 Leu by the S1 subsite is the key interaction. This is of particular interest because it suggests that peptidomimetic compounds, preferred over peptides as drug molecules due to better pharmacokinetic properties, would be well tolerated.

To address the question of whether the inhibitors identified would function in cells, primary human foreskin fibroblasts (HFFs) were pretreated with the compounds and subsequent intoxication by recombinant TcdB holotoxin quantified by measuring the degree of cytopathosis (observed as cell rounding). The order of potency matched the ranking by IC_{50}, with the most potent inhibitor, Hpa-SerLeu-AOMK, exhibiting a cellular IC_{50} of 20 μM. Further analysis showed that inhibiting TcdB CPD directly affected the holotoxin effector domain activity. This was achieved by following the glucosylation of the holotoxin effector domain's cellular target, a Rho GTPase called Rac1. This was required because AOMK inhibitors with Leu in the P1 position have been shown to have some weak cross-reactivity with cathepsins. However, the use of JPM-OEt, a cell-permeable cathepsin inhibitor [17, 18], as a control had no effect on holotoxin effector domain function.

A key event in the function of *C. difficile* glucosylating toxins is activation of the CPD by eukaryotic-specific inositol hexakisphosphate ($InsP_6$). To facilitate the study of this toxin activation in more detail, fluorescent and biotin-tagged analogues of the SerLeu-AOMK inhibitor were designed. Both of these singly labelled probes were seen to covalently modify recombinant TcdB(1–804) only in the presence of $InsP_6$. Importantly, the catalytically inactive TcdB(1–804)C698A (Cys698 replaced by Ala) mutant was not labelled, regardless of the presence of $InsP_6$, suggesting the specificity of the probe attachment mechanism. However, the cleavage product TcdB(544–804) was more prone to probe labelling, presumably due to higher reactivity and/or greater accessibility.

By employing the cell-permeable Cy5-functionalised probe AWP19 (Fig. 3), off-target effects could be identified. In both primary HFF and RAW macrophage cells, the only other target of AWP19 found was cathepsin B. However, the control compound cathepsin inhibitor JPM-OEt (Fig. 3) had no effect on cell rounding or Rac1 glucosylation upon toxin addition, confirming that inhibition of CPD function was solely responsible for the observed reduction in toxin-induced cytopathosis. The probes therefore provide tools for direct measurement of allosteric toxin activation by binding to $InsP_6$ in biochemical and cell-based models.

More recently, the same authors published a study into the structure and precise molecular mechanism of the allosteric circuit, by which it was shown that specific residues in a β-hairpin form a link between the $InsP_6$ binding site and the CPD active site [19]. These residues undergo a highly orchestrated series of interactions and structural rearrangements to align catalytic residues and permit $InsP_6$-mediated activation of the CPD. Mutational analyses were performed, revealing that 12 out of the 253 CPD residues are involved in $InsP_6$ signalling. The activity-based probe AWP19 played a key role as the read-out for TcdB CPD activity during investigation of the mechanisms of signal transduction.

2.3.2 Proteases in *S. aureus*

S. aureus, is a facultative anaerobic, Gram-positive coccus. Approximately 20% of the human population are long-term carriers of *S. aureus*, whilst some 60% are intermittent carriers [20]. This bacterium has increased in significance recently due to the appearance of methicillin-resistant strains (i.e., MRSA). As opposed to targeting viability, β-lactones have been shown to target virulence, a key component of which is the caseinolytic protease (ClpP). This enzyme was amongst a collection of proteins isolated using a panel of β-lactones in "fishing expeditions" in several Gram-positive and Gram-negative bacteria. These included enzymes involved in primary and secondary metabolism, nucleotide synthesis, antibiotic

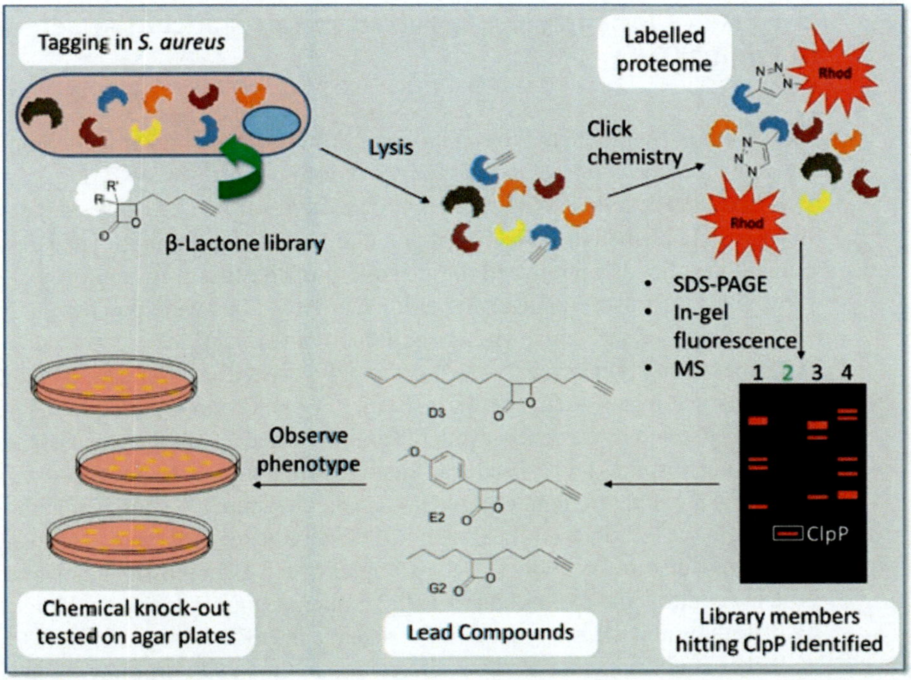

Fig. 4 Identification of alkyne-tagged β-lactone probes that target ClpP in *S. aureus* using click-chemistry, in-gel fluorescence and mass spectrometry. Subsequently, lead compounds were administered to *S. aureus* and changes to the global bacterial proteome monitored. Particularly, changes to the levels of haemolysis and proteolysis enzymes (haemolysins, proteases, lipases and nucleases) were observed

resistance and cell-wall biosynthesis, virulence, detoxification and others of, as yet, unknown function. Enzyme classes represented in this study included ligases, oxidoreductases, hydrolases and transferases [21].

Sieber et al. applied synthetic *trans*-β-lactone ABPs to the study of ClpP in *S. aureus* and showed that targeting this virulence-associated enzyme offers an attractive means to attenuate the harmful effects of infection, potentially allowing the host's immune system to deal with the bacteria [22]. Although ΔClpP mutant strains of *S. aureus* have shown the key contribution of this enzyme to virulence regulation [23], effective ClpP inhibitors had yet to be reported. However, synthetic *trans*-β-lactones were found to inhibit ClpP and dramatically decrease the expression of virulence factors such as haemolysins, proteases, DNases and lipases, which suppress host immune response and induce tissue inflammation and necrosis (Fig. 4). Building on earlier work using β-lactones as privileged structures for labelling enzyme classes in various bacteria [21], cell-permeable and highly ClpP-specific *trans*-β-lactone ABPs were used to monitor the activity of this protease in *S. aureus* and MRSA and quantify the inhibition.

3 Probes for Enzymes Involved in Resistance: β-Lactams and MRSA

First prepared synthetically by Staudinger in 1907 [24], and gaining clinical significance with the isolation of penicillin by Chain and Florey in 1940 [25], the β-lactam structure has remained at the core of antibiotics research. Ironically, in the same year that Chain published the isolation of penicillin and before penicillin had even entered clinical use, he also published the discovery of the first β-lactamase, penicillinase [26]. Since then, β-lactamase production has spread to a wide population of bacteria that previously either only synthesised it in minute quantities or not at all. However, ABPs based on the structure of the most widely resisted antibiotic are a natural choice for investigating mechanisms of resistance. For the most part, MRSA infections cannot be treated by currently available antibiotics. These strains have spread widely throughout hospitals and caused considerable problems for the healthcare community and the general public, rendering MRSA a quintessential "hospital superbug". Such bacteria make use of several pathways to develop resistance, spread across complex signal transduction networks. For example, MecR1 (a sensor/signal transduction protein) controls the expression of PBP2′, a penicillin-binding protein not inhibited by β-lactams and the major cause of resistance in MRSA.

Sieber et al. were able to show differences in the in situ profiling of non-resistant and resistant strains of *S. aureus* (SA and MRSA, respectively) [2]. Specifically, by labelling of the MecR1 antibiotic sensor domain with probes resembling natural antibiotics and using an artificial β-lactam probe, PBP2′,they were able to identify two previously uncharacterised MRSA enzymes that displayed the ability to hydrolyse β-lactam antibiotics from the penicillin family. One of these, a metallo-β-lactamase, belongs to a novel class of enzymes of increasing interest in understanding mechanisms of resistance and for which few chemical tools exist for further study.

In order to maximise the diversity of targets hit, three probes were based on the scaffolds from ampicillin, cephalosporin and aztreonam, and seven probes were intended to label enzymes other than penicillin-binding proteins (PBPs) and were reactive to active site nucleophiles Cys and Ser. Labelling was carried out on intra- and extracellular proteomes, with the post-lysis membrane fraction showing the greatest differentiation between the SA and MRSA strains. In addition to an uncharacterised hydrolase (Hy30), a metabolic enzyme involved in fatty acid biosynthesis (KASIII) and an ATP-dependent virulence regulator (ClpP), similar to those found in previous β-lactam profiling experiments with antibiotic-sensitive bacteria, these studies revealed several enzymes present only in the MRSA strain. These included PBP2 and PBP2′ (the amide-activated compound NCO (Fig. 5) was presumably able to overcome the low reactivity of these compounds towards β-lactams) a serine protease (SPD_0), a carboxyesterase precursor-like protein (E28) and a dipeptidase (Dipep). Interestingly, no targets were observed in SA that were not found in MRSA.

As introduced above, the antibiotic sensing domain of MecR1 detects β-lactams in the extracellular environment and induces expression of PBP2′, which takes over

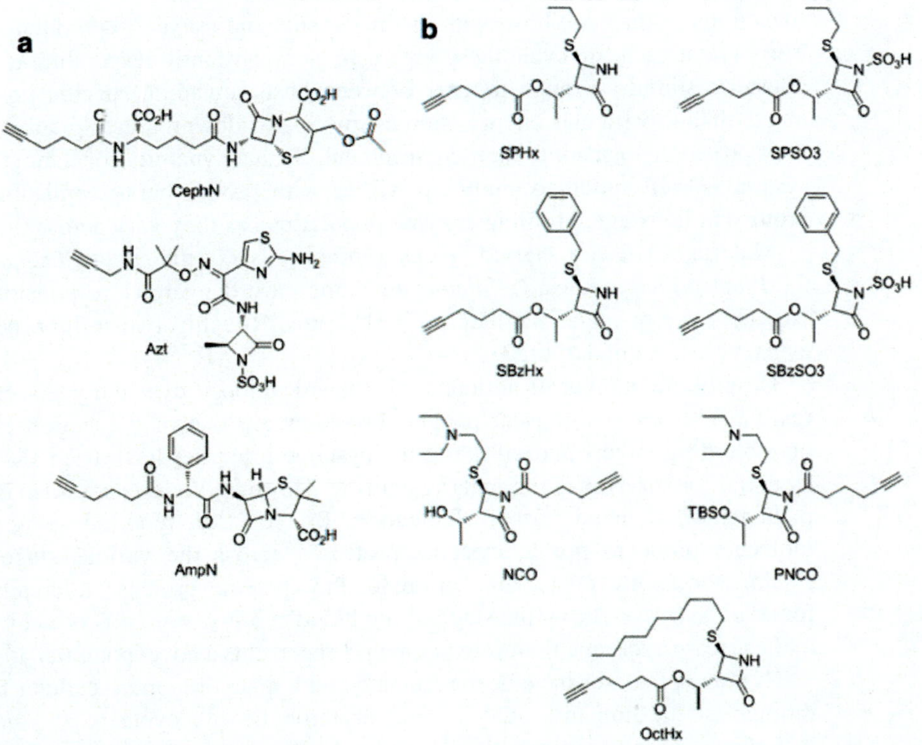

Fig. 5 Structures of (**a**) natural and (**b**) synthetic β-lactam probes

cell wall biosynthesis capacity from PBPs inhibited by antibiotics. The natural β-lactam probes (CephN, Azt and AmpN, Fig. 5) detected this domain (Chain A, Mecr1 unbound extracellular antibiotic-sensor domain MECR1) in the extracellular proteome of MRSA (but not in SA). As well as β-lactamases (e.g. E28), the identification of dipeptidase PepV (Dipep), a member of the less well-studied metallo-β-lactamases, alludes to an increasingly significant problem in combating resistance as these enzymes are unaffected by current β-lactamase inhibitors.

Altogether, this study proved an excellent demonstration of how these chemical tools can be used to monitor the activity and function of enzymes known to be involved in resistance, as well as facilitate the identification of previously uncharacterised enzymes, and is undoubtedly a significant contribution to the search for future therapeutic interventions.

4 Protozoan Parasites: Malaria and Falcipain

As opposed to the prokaryotic bacteria, protozoa are a diverse set of eukaryotic organisms, often with complex life cycles. These entities exist in various forms, from proliferative trophozoites through to dormant cysts, and some spending their lives in more than one host, with distinct sexual and asexual reproduction phases. Throughout each life cycle these organisms are constantly responding to different allosteric stimuli, through transfer between hosts, invading specific host tissues, metabolising particular energy sources, by up- or downregulating subsections of their proteome to adapt to their environment. Such a dynamic, constantly changing system is well suited to study by ABPP, with probes being applicable to any protozoan life stage, profiling enzyme populations as they vary across the cycle.

Malaria is a disease caused by eukaryotic protists of the genus *Plasmodium*. Of the *Plasmodium* species, *P. falciparum* is the most dangerous, responsible for the vast majority of malaria infections and deaths. Roughly two million people die every year from malaria [27].

Despite the number of antimalarial drugs developed over the years, multidrug resistance is an ever-present danger. Proteases are one of the largest groups of possible drug targets and, of these, the cysteine proteases have been shown to be essential for survival of the malaria parasite. However, their exact roles in disease pathogenesis demand further elucidation. Bogyo et al. reported using a small-molecule probe to profile cysteine proteases across the various stages of the *P. falciparum* life cycle. One protease, the cysteine protease falcipain 1, was found to be active during the stage of red blood cell invasion; this is a key moment in the disease cycle and therefore falcipain 1 represents a novel potential drug target.

Cysteine proteases have been associated with many important cellular functions throughout the human-hosted asexual erythrocytic life cycle of *P. falciparum*, including cell invasion, but also red blood cell rupture via cleavage of ankyrin, subsequent parasite release from the parasitophorous vacuole and haemoglobin degradation. The authors profiled cysteine protease activity across four stages of the parasite, namely ring, trophozoite, schizont and merozoite, and looked at the soluble vs. insoluble fractions. Merozoites formed during the blood stage of the life cycle are released from the liver to invade erythrocytes. Over a period of 48 h the internalised parasites transform into the ring form, then the trophozoite form (to metabolise haemoglobin) and then they replicate in the schizont form to create more merozoites, which are released by cell rupture to repeat the cycle. Merozoites must quickly reinvade a fresh erythrocyte, or become nonviable.

Bogyo et al. used a radioisotope-containing cysteine protease probe (^{125}I-DCG-04) for simple gel-image based profiling, and a biotin affinity tag version of DCG-04 for enrichment and subsequent MS-based characterisation [28]. All cysteine proteases identified were of the papain family, as expected from the E64-type warhead used. In the detergent insoluble fractions of the four parasite forms only falcipain 1 was identified, implying a distinct role in localisation. Soluble fraction falcipain 1 was observed only in the late schizont and merozoite forms. Falcipain 1 was the first cysteine protease gene cloned from *P. falciparum*, but recombinant expression has remained a challenge, such that there has been little progress in its biochemical characterisation and functional assignment. However, this renders it a good subject for ABPP studies. In addition, falcipains 2 and 3 were isolated from the soluble fractions of the trophozoite and schizont forms where they are known to be a feature of the food vacuole and effect haemoglobin degradation. Another protein identified in very small quantities from detergent soluble and insoluble fractions was human calpain 1, which the authors claim not to result from erythrocyte contamination as these were lysed prior to isolation of proteases with the probe; however this is not discussed further. Furthermore, cathepsin C-like protease was also identified in the soluble fractions of the trophozoite and schizont forms, but no biological function has been reported for this enzyme.

Heightened falcipain 1 activity during the merozoite stage implied a role in either erythrocyte invasion or rupture (or both). Further detailed imaging studies of highly synchronised merozoites were carried out using an anti-falcipain 1 antibody. These showed that falcipain 1 was localised to apical organelles, which facilitate host cell invasion, during which their contents are secreted. Their presence in organelles discreet from those occupied by rhoptry protein and micronemal proteins (RopH2 and EBA-175 respectively, markers for two invasion-specific organelles in merozoites) suggests a third set of apical entities, or a new subset, implying a greater assortment of invasion machinery than first thought. These data clearly point to falcipain 1 having a key role in merozoite invasion.

The techniques required for gene knockout in *P. falciparum* are not yet sufficiently developed for phenotypic functional characterisation of falcipain 1. Instead the authors looked for a chemical genetic approach, using a falcipain 1-specific inhibitor. By varying two out of three residues in a three amino acid sequence attached to the warhead, many hundreds of compounds were generated. Incubation of these with lysates, followed by cysteine protease activity profiling with the ^{125}I-DCG-04 probe allowed the identification of compounds with falcipain 1-specific activity. The most selective cysteine protease inhibitor, YA29-Eps(S,S), showed a 25-fold preference for falcipain 1. Parasites were treated with either papain family protease inhibitor E64d, falcipain 1 specific YA29-Eps(S,S) or inactive YAG-Eps (S,S). The inactive control showed no effect, but the papain inhibitor E64d and falcipain 1 inhibitor both showed a dose dependent reduction in new ring-stage parasites post-reinvasion. But whilst treatment with high concentrations of E64d also resulted in an enlargement of the food vacuole (consistent with previous reports), the same was not observed for YA29-Eps(S,S), allowing development into schizonts and rupture to release new merozoites. This confirmed that falcipain 1 was not involved in haemoglobin degradation or erythrocyte rupture post-schizont, but must be important during extracellular merozoite invasion.

This work elegantly demonstrates power of combined ABPP and chemical genetics approaches in understanding microbial pathogenesis, and has made a very real contribution to the identification of potential novel drug targets.

5 Quorum Sensing

For microorganism communities, energy-demanding phenomena such as biofilm formation, virulence factor expression and even bioluminescence are only beneficial once the population density has passed a certain threshold [29]. Bacteria

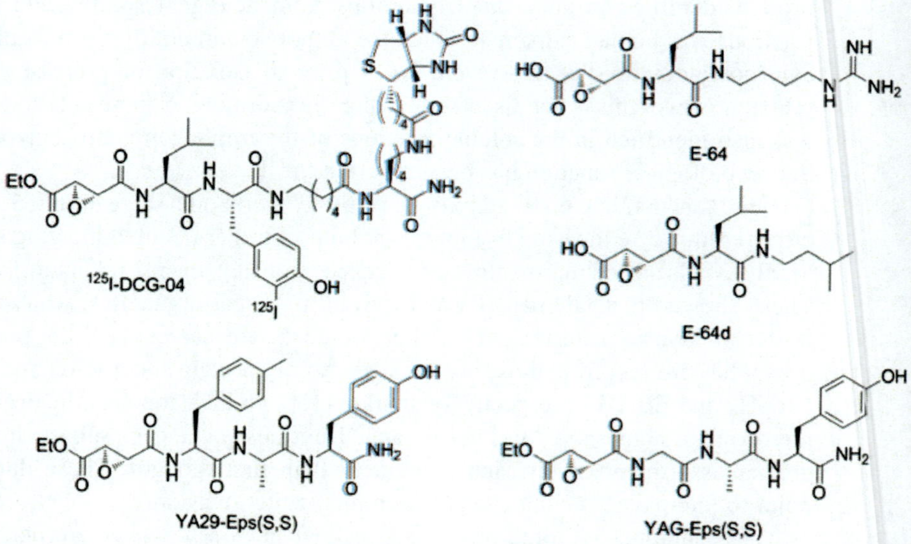

Fig. 6 Structures of epoxide-based inhibitors and activity-based probes used in the study of falcipain 1

(amongst other organisms) have evolved a cell-to-cell communication mechanism known as "quorum sensing" to synchronise their efforts upon the population passing a certain density and becoming "quorate" [30]. Dependent on the constant release of signalling molecules and their subsequent receptor-based recognition by neighbouring organisms, quorum sensing is an interesting target for antimicrobial therapeutics. Knowledge of the mechanisms by which quorum sensing occurs is increasing steadily, with the diffusion of *N*-acyl homoserine lactones (AHLs) being a key factor in many Gram-negative bacteria [6]. Conversely, some bacteria have AHL-degrading processes, which quench quorum sensing in competitor organisms as a means of competition [31].

Designed to regulate gene expression amongst their own populace, bacterial AHL autoinducers have also been shown to influence eukaryotic cells. For example, *Pseudomonas aeruginosa* inducer *N*-(3-oxododecanoyl)homoserine lactone (3-oxo-C_{12}-HSL, C12, Fig. 6) has been shown to effect cytokine production (e.g. interferon-γ) by immune cells both in vitro and in vivo [32]. More recently, a mechanism was reported by which C12 disrupts NF-$\kappa\beta$ signalling in activated mammalian cells, implying a deliberate downregulation of host immune response [33]. Despite the significance of AHLs in these and other interorganism processes, the interactions between C12 and its receptors are poorly understood. To provide elucidation, Meijler et al. designed ABPs based on the structure of AHL C12 (Fig. 7) and used them in the identification and validation of quorum sensing receptors [34]. The probe design avoided altering the lactone ring, such modifications being poorly tolerated, but focussed instead on making minimal alterations to the 9-carbon alkyl chain. The strategy involved two stages of ligation:in the first instance, a photoreactive diazirine was used to covalently attach the C12 analogue to the receptor, and in the second, an alkyne located at the alkyl terminus was intended to affix a reporter tag to the receptor via the probe. Despite the amount of ring strain, diazirines possess excellent chemical stability. Upon exposure to light, they eject a molecule of nitrogen to produce a highly reactive carbine, which reacts readily with neighbouring chemistry to label, in the case of biomolecules, site-specifically [35].

The dual-function probe was shown to mimic C12, activating the LasI–LasR quorum sensing system and inducing bioluminescence in a reporter strain of *E. coli* (pSB1075) [36]. In a *P. aeruginosa* LasI mutant lacking the ability to synthesise C12, the probe induced elastase production, a virulence factor under Las control. However, it should be noted that although native C12 displayed an IC_{50} of 1 μM, the probe was significantly more active with an IC_{50} of 30 nM. The apparent activity was reversed in macrophages, with induction of initiator factor eIF2α slightly less with the probe than with C12, but effective nonetheless. The probe and C12 showed comparable inhibition of germ tube formation in *Candida albicans*. It was thus determined that the probe recapitulated the activity of C12 over a range of prokaryotes and eukaryotes.

The authors demonstrated their probe's ability to label recombinant LasR ligand binding domain (LasR-LBD) expressed in *E. coli*, detecting labelled protein by mass spectrometry but without going on to utilise the alkyne for click chemistry. However, in a follow-up report [37], the diazirine and alkyne design had been surpassed by a version that covalently attached to the receptor via an isothiocyanate (previously reported in C12-based quorum sensing inhibitors, reacting specifically with the Cys79 in LasR-LBD [38]) and to which an aminooxy-BODIPY could be attached directly via the β-carbonyl group (Fig. 7). Interestingly, Janda et al. reported the synthesis of both alkyne- and azide-containing C12 analogues as quorum sensing probes to investigate mammalian targets of C12. They found, quite unexpectedly, that only the azide-containing analogue displayed comparable activity in C12-induced signalling events in bone marrow-derived macrophages, whereas the alkyne's activity was greatly diminished [39].

Ordinarily, the ligation between a ketone and an aminooxy species is carried out under acidic conditions ($pH < 4.5$) because the reaction can take days to complete at neutral pH. However, the application of aniline catalysis has rendered neutral ligation at very good rates possible [40]. In this way the labelling of recombinant LasR-LBD was demonstrated, and was similar to that of the diazirine probe shown in Fig. 7, but with progression to ligation of a BODIPY fluorophore. Furthermore, treatment of *P. aeruginosa* with probe and ensuing in vitro BODIPY ligation on whole cell lysates allowed visualisation of LasR by in-gel fluorescence. Given the low expression level of LasR and its poor stability this demonstrated the extremely high sensitivity of the technique, and also showed the high degree of chemoselectivity of the BODIPY–oxime since only a very small amount of non-specific labelling was observed.

This bioorthogonality was exploited to achieve in vivo labelling in both *E. coli* overexpressing LasR-LBD and *P. aeruginosa*. The bacteria were fed probe for 12 h before washing to remove unused lactone. The cell-permeable BODIPY-oxime was then fed to the cells for a further 12 h at pH 6.6 at 8 °C, after which the bacteria

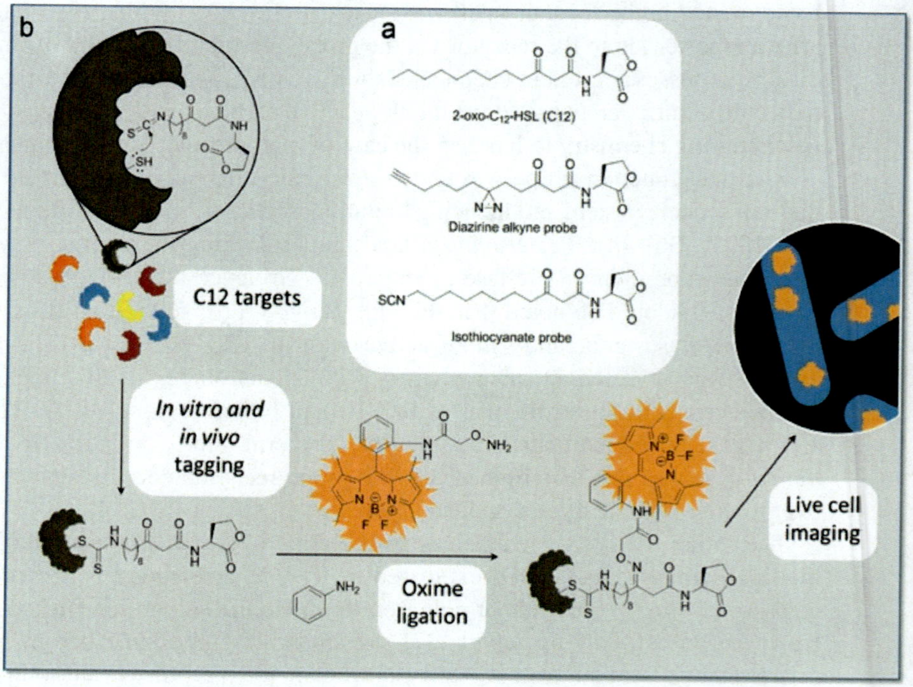

Fig. 7 (**a**) Structures of 2-oxo-C_{12}-HSL (C12), diazirine–alkyne probe and isothiocyanate probes. (**b**) ABPP strategy using tagging via the –SCN group followed by aniline-catalysed oxime ligation to achieve live bacterial cell imaging

were washed, lysed and prepared for SDS–PAGE. Labelling was observed in *E. coli* and could be undone by the addition of C12 during probe feeding, which outcompeted the probe and demonstrated selectivity.

In *P. aeruginosa*, despite considerably lower levels of LasR when compared to the *E. coli* system, the same protocol resulted in detection of a band by in-gel fluorescence, which could be excised and the identification of LasR achieved by proteomic MS methods with 32% total sequence coverage. The methodology was also extended to allow analysis by flow cytometry. In this way, the authors were able to demonstrate that cell density had an effect on LasR labelling. As expected, fluorescence increased with cell density up to the point at which quorum was achieved (around $OD_{600} = 6$), beyond which fluorescence began to weaken.

The widely accepted view is that *P. aeruginosa* LasR is membrane bound via its N-terminus and upon complex formation with C12 a structural change occurs that allows binding to the target DNA. However, whether the LasR remains membrane bound throughout this process, or indeed whether LasR is subject to subcellular localisation or is evenly distributed throughout the cell, remains unknown. Having established the in vitro labelling methodology in *P. aeruginosa*, a method for whole-cell imaging was developed. It was observed that fluorescence was localised to the polar regions, in accordance with earlier observations that an EGFP-fusion protein of LasR-related quorum sensing regulator RhlR was similarly distributed

[41]. It was also noted that the intensity was dependent on cell density, as recorded by the earlier flow cytometry work. As the concentration of probe was increased, the degree of membrane association was observed to reduce, possibly providing some evidence that the LasR dissociates upon binding C12; however, much higher resolution microscopy is needed to confirm these conclusions.

Despite the preliminary nature of the imaging results, the demonstration of the power of the isothiocyanate ABP and the attendant oxime ligation is highly significant, and the degree of selectivity observed in the aniline-catalysed ligation is remarkable.

6 Summary

We have described here a number of approaches to the design and utilisation of activity-based protein profiling for unravelling the highly complex and dynamic systems involved in microbial pathogenesis. Whether ABPs have been used for answering some of the many fundamental questions regarding microbe biology, trying to fully understand mechanisms of resistance, or inventing entirely new therapeutic strategies, they have more than shown their worth as innovative and flexible tools in this field. Combining the inhibitor design and synthesis skills of medicinal chemistry with the latest molecular biology techniques, the field of chemical proteomics is rapidly emerging as one of the foremost disciplines in the effort to advance post-genomic life sciences. As well as providing a particularly stimulating and rewarding endeavour for those groups involved, it also serves as a model for the type of cross-disciplinary research that is increasingly characterising modern science.

References

1. World Health Organization (2011) WHO, Geneva. http://www.euro.who.int/en/home
2. Staub I, Sieber SA (2009) Beta-lactam probes as selective chemical-proteomic tools for the identification and functional characterization of resistance associated enzymes in MRSA. J Am Chem Soc 131:6271
3. Heal WP, Dang TH, Tate EW (2011) Activity-based probes: discovering new biology and new drug targets. Chem Soc Rev 40:246
4. Heal WP, Wickramasinghe SR, Tate EW (2008) Activity based chemical proteomics: profiling proteases as drug targets. Curr Drug Discov Technol 5:200
5. Tang C, Holden D (1999) Pathogen virulence genes–implications for vaccines and drug therapy. Br Med Bull 55:387
6. Miller MB, Bassler BL (2001) Quorum sensing in bacteria. Annu Rev Microbiol 55:165
7. Masuda K, Itoh M, Kawata T (1989) Characterization and reassembly of a regular array in the cell wall of *Clostridium difficile* GAI 4131. Microbiol Immunol 33:287
8. Sleytr UB, Messner P (1983) Crystalline surface layers on bacteria. Annu Rev Microbiol 37:311

9. Dang TH, de la Riva L, Fagan RP, Storck EM, Heal WP, Janoir C, Fairweather NF, Tate EW (2010) Chemical probes of surface layer biogenesis in *Clostridium difficile*. ACS Chem Biol 5:279
10. Schirmeister T, Klockow A (2003) Cysteine protease inhibitors containing small rings. Mini Rev Med Chem 3:585
11. Böttcher T, Sieber SA (2010) Showdomycin as a versatile chemical tool for the detection of pathogenesis-associated enzymes in bacteria. J Am Chem Soc 132:6964
12. Uehara Y, Fisher JM, Rabinovitz M (1980) Showdomycin and its reactive moiety, maleimide. A comparison in selective toxicity and mechanism of action in vitro. Biochem Pharmacol 29:2199
13. Rupnik M, Wilcox MH, Gerding DN (2009) *Clostridium difficile* infection: new developments in epidemiology and pathogenesis. Nat Rev Microbiol 7:526
14. Lyras D, O'Connor JR, Howarth PM, Sambol SP, Carter GP, Phumoonna T, Poon R, Adams V, Vedantam G, Johnson S, Gerding DN, Rood JI (2009) Toxin B is essential for virulence of *Clostridium difficile*. Nature 458:1176
15. Puri AW, Lupardus PJ, Deu E, Albrow VE, Garcia KC, Bogyo M, Shen A (2010) Rational design of inhibitors and activity-based probes targeting *Clostridium difficile* virulence factor TcdB. Chem Biol 17:1201
16. Shen A, Lupardus PJ, Albrow VE, Guzzetta A, Powers JC, Garcia KC, Bogyo M (2009) Mechanistic and structural insights into the proteolytic activation of *Vibrio cholerae* MARTX toxin. Nat Chem Biol 5:469
17. Greenbaum D, Medzihradszky KF, Burlingame A, Bogyo M (2000) Epoxide electrophiles as activity-dependent cysteine protease profiling and discovery tools. Chem Biol 7:569
18. Bogyo M, Verhelst S, Bellingard-Dubouchaud V, Toba S, Greenbaum D (2000) Selective targeting of lysosomal cysteine proteases with radiolabeled electrophilic substrate analogs. Chem Biol 7:27
19. Shen A, Lupardus PJ, Gersch MM, Puri AW, Albrow VE, Garcia KC, Bogyo M (2011) Defining an allosteric circuit in the cysteine protease domain of *Clostridium difficile* toxins. Nat Struct Mol Biol 18:364
20. Kluytmans J, van Belkum A, Verbrugh H (1997) Nasal carriage of *Staphylococcus aureus*: epidemiology, underlying mechanisms, and associated risks. Clin Microbiol Rev 10:505
21. Böttcher T, Sieber SA (2008) Beta-lactones as privileged structures for the active-site labeling of versatile bacterial enzyme classes. Angew Chem Int Ed Engl 47:4600
22. Böttcher T, Sieber SA (2008) Beta-lactones as specific inhibitors of ClpP attenuate the production of extracellular virulence factors of *Staphylococcus aureus*. J Am Chem Soc 130:14400
23. Frees D, Qazi SN, Hill PJ, Ingmer H (2003) Alternative roles of ClpX and ClpP in *Staphylococcus aureus* stress tolerance and virulence. Mol Microbiol 48:1565
24. Staudinger H (1907) Zur Kenntniss der Ketene. Diphenylketen. Justus Liebigs Ann Chem 356:51
25. Chain E, Florey HW, Gardner AD, Heatley NG, Jennings MA, Orr-Ewing J, Sanders AG (1940) Penicillin as a chemotherapoutic agent. Lancet 236:226
26. Abraham EP, Chain E (1940) An enzyme from bacteria able to destroy penicillin. Nature 46:837
27. Breman JG (2001) The ears of the hippopotamus: manifestations, determinants, and estimates of the malaria burden. Am J Trop Med Hyg 64:1
28. Greenbaum DC, Baruch A, Grainger M, Bozdech Z, Medzihradszky KF, Engel J, DeRisi J, Holder AA, Bogyo M (2002) A role for the protease falcipain 1 in host cell invasion by the human malaria parasite. Science 298:2002
29. Hastings JW, Nealson KH (1977) Bacterial bioluminescence. Annu Rev Microbiol 31:549
30. Engebrecht J, Nealson K, Silverman M (1983) Bacterial bioluminescence: isolation and genetic analysis of functions from *Vibrio fischeri*. Cell 32:773

31. Chan KG, Atkinson S, Mathee K, Sam CK, Chhabra SR, Camara M, Koh CL, Williams P (2011) Characterization of N-acylhomoserine lactone-degrading bacteria associated with the *Zingiber officinale* (ginger) rhizosphere: co-existence of quorum quenching and quorum sensing in Acinetobacter and Burkholderia. BMC Microbiol 11:51
32. Ritchie AJ, Jansson A, Stallberg J, Nilsson P, Lysaght P, Cooley MA (2005) The *Pseudomonas aeruginosa* quorum-sensing molecule N-3-(oxododecanoyl)-L-homoserine lactone inhibits T-cell differentiation and cytokine production by a mechanism involving an early step in T-cell activation. Infect Immun 73:1648
33. Kravchenko VV, Kaufmann GF, Mathison JC, Scott DA, Katz AZ, Grauer DC, Lehmann M, Meijler MM, Janda KD, Ulevitch RJ (2008) Modulation of gene expression via disruption of NF-kappaB signaling by a bacterial small molecule. Science 321:259
34. Dubinsky L, Jarosz LM, Amara N, Krief P, Kravchenko VV, Krom BP, Meijler MM (2009) Synthesis and validation of a probe to identify quorum sensing receptors. Chem Commun 47:7378
35. Blencowe A, Hayes W (2005) Development and application of diazirines in biological and synthetic macromolecular systems. Soft Matter 1:178
36. Winson MK, Swift S, Fish L, Throup JP, Jorgensen F, Chhabra SR, Bycroft BW, Williams P, Stewart GS (1998) Construction and analysis of luxCDABE-based plasmid sensors for investigating N-acyl homoserine lactone-mediated quorum sensing. FEMS Microbiol Lett 163:185
37. Rayo J, Amara N, Krief P, Meijler MM (2011) Live cell labeling of native intracellular bacterial receptors using aniline-catalyzed oxime ligation. J Am Chem Soc 133:7469
38. Amara N, Mashiach R, Amar D, Krief P, Spieser SA, Bottomley MJ, Aharoni A, Meijler MM (2009) Covalent inhibition of bacterial quorum sensing. J Am Chem Soc 131:10610
39. Garner AL, Yu J, Struss AK, Lowery CA, Zhu J, Kim SK, Park J, Mayorov AV, Kaufmann GF, Kravchenko VV, Janda KD (2011) Synthesis of 'clickable' acylhomoserine lactone quorum sensing probes: unanticipated effects on mammalian cell activation. Bioorg Med Chem Lett 21:2702
40. Dirksen A, Hackeng TM, Dawson PE (2006) Nucleophilic catalysis of oxime ligation. Angew Chem Int Ed Engl 45:7581
41. Ventre I, Ledgham F, Prima V, Lazdunski A, Foglino M, Sturgis JN (2003) Dimerization of the quorum sensing regulator RhlR: development of a method using EGFP fluorescence anisotropy. Mol Microbiol 48:187

Top Curr Chem (2012) 324: 137–162
DOI: 10.1007/128_2011_284
© Springer-Verlag Berlin Heidelberg 2011
Published online: 15 November 2011

Functional Analysis of Protein Targets by Metabolomic Approaches

Yun-Gon Kim and Alan Saghatelian

Abstract Proteomics methods, such as activity-based protein profiling, can be used to connect proteins to biology and disease. Some proteins found through unbiased methods are not well characterized, which makes it difficult to ascertain the role of these proteins. Metabolomics approaches are useful in characterizing proteins that regulate or bind metabolites. Here, we provide examples of the development and use of metabolomics approaches to elucidate protein–metabolite interactions.

Keywords Enzymes · Liquid chromatography-mass spectrometry (LC-MS) · Metabolite binding proteins · Metabolomics

Contents

Y.-G. Kim and A. Saghatelian (✉)
Department of Chemistry and Chemical Biology, Harvard University, 12 Oxford St., Cambridge, MA, USA
e-mail: saghatelian@chemistry.harvard.edu

Abbreviations

15d-PGJ2	15-Deoxy-Δ-12,14-prostaglandin-J2
ABPP	Activity-based protein profiling
ANS	1-Anilinonaphthalene-8-sulfonate
CRABP2	Cytosolic retinoic acid binding protein 2
DBD	DNA-binding domain
DMP	Discovery metabolite profiling
DPP4	Dipeptidyl peptidase 4
FAAH	Fatty acid amide hydrolase
FTMS	Fourier transform mass spectrometry
GST	Glutathione *S*-transferase
IDH	Isocitrate dehydrogenase
IDMS	Isotope dilution mass spectrometry
LBD	Ligand-binding domain
LC-MS	Liquid chromatography–mass spectrometry
LPA	Oleyl lysophosphatidic acid
MAGE	Monoalkylglycerol ether
MAGL	Monoacylglycerol lipase
NAE	*N*-acyl ethanolamine
NAT	*N*-acyl taurine
NMR	Nuclear magnetic resonance
PHI	Peptide histidine isoleucine
PMI	Protein–metabolite interactions
PPAR	Peroxisome proliferator-activated receptor
R132H	Arginine 132 switched to histidine
StarD3	StAR-related lipid transfer domain 3
THC	Tetrahydrocannabinol

1 Introduction

1.1 The Need for Metabolomics as a Functional Proteomics Tool

Genes encode RNA, and RNA encodes proteins, which then carry out most of the biochemistry that runs cells and organisms [1]. A number of these proteins have roles in shaping the metabolome, the complete collection of metabolites in a cell, tissue, or organism [2]. Metabolites provide cells with energy, are building blocks for macromolecules and cellular structures, and can also act as signaling molecules [1]. The types of metabolites and their concentrations are controlled through interactions between the proteome and the metabolome [3, 4]. Identifying these connections is important in the functional characterization of proteins, and in understanding how the metabolome is regulated [5].

The rise of global analysis methods for nucleic acids [6–8] and proteins [9–11] demonstrates the value of unbiased analysis in molecular biology. Activity-based protein profiling (ABPP), for example, measures changes in enzyme activity between biological samples. This method has been used to discover enzymes that are associated with disease processes, such as cancer cell invasiveness [12, 13]. Similarly, unbiased analysis of the metabolome led to the development of metabolomics approaches [5, 14–19].

The scientific basis of metabolomics can be traced to early experiments in medicine by Garrod, who discovered alkaptonuria, the first inborn error of metabolism, based on changes in urine metabolites [20, 21]. In Garrod's work, a build-up of homogentisic acid in the urine of patients with alkaptonuria caused the urine to turn black upon standing due to oxidation of homogentisic acid. This color change provided a metabolic readout that allowed Garrod to determine that there was a change in the underlying metabolism of his patients. Later studies revealed that patients suffering from alkaptonuria are deficient in the homogentisic acid catabolic enzyme, homogentisate-1,2-dioxygenase, which explains the phenotype [22]. Today, metabolomics research continues to correlate metabolic differences to phenotypic changes [23].

The identification of changes in the metabolome can help characterize the biochemical functions of enzymes in the proteome [5, 24]. Not all metabolic changes, however, provide easy readouts. Therefore, more sophisticated analytical methods have been developed to detect and quantify changes in metabolome. The application of modern analytical tools, such as nuclear magnetic resonance (NMR) [19] and mass spectrometry (MS) [5, 24], are the primary tools of metabolomics researchers. In particular, MS has found increased usage as mass spectrometers have improved their sensitivity and mass resolution.

1.2 The Metabolome

The metabolome refers to the entire collection of metabolites, including lipids, sugars, amino acids, and nucleosides within an organism [2]. Though there is still some debate as to the exact number of metabolites, the current estimate is ~6,800 human metabolites [25]. Compared to large biopolymers consisting of thousands of atoms, such as proteins and nucleic acids, the structures of many metabolites seem simple. However, this simplicity masks the unique challenge associated with analysis of the metabolome due to the distinct physicochemical properties of different classes of metabolites. For example, the isolation and analysis protocols

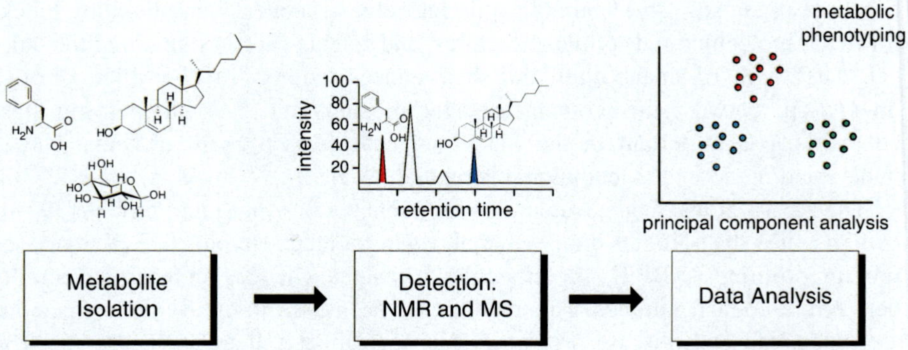

Fig. 1 The typical metabolomics workflow has three key steps: the isolation of metabolites, detection of the metabolites, and data analysis. The isolation step is typically determined by the class of metabolite being measured because of the physicochemical properties of different metabolite classes (i.e., hydrophobic, hydrophilic), which require different enrichment protocols. Two principle methods for metabolite detection are NMR- and MS-based methods. Finally, the data analysis can be performed in a variety of ways depending on the problem

necessary for the analysis of lipids [26] provide no coverage for hydrophilic molecules (e.g., sugars, amino acids) [27, 28]. Accordingly, there are specific protocols for the analysis of hydrophobic [5, 24, 29] and hydrophilic [30] metabolites that provide full coverage across the metabolome (Fig. 1).

1.3 NMR Metabolomics

We will briefly highlight some examples using NMR methods [14, 19] but the remainder of this chapter will focus on MS and liquid chromatography–mass spectrometry (LC-MS) metabolomics [5, 15, 18]. We should also point out that there are a number of examples of both NMR and MS methods being used in the discovery of medicinally important biomarkers [23, 31]; however, this review will focus more on the use of metabolomics to characterize proteins.

Raamsdonk and colleagues set out to develop a metabolomics method that could be used to characterize proteins of unknown function in yeast [14]. Using an NMR approach they analyzed intracellular metabolites in mutants of *Saccharomyces cerevisiae*. The resulting NMR spectra were then analyzed by multivariate analysis, including principle component analysis (PCA) to identify differences in the spectra that can distinguish different mutants (Fig. 2). Two important results came out of these studies that would reveal the value of metabolomics in biological research.

First, none of the mutants they were studying (pfk26Δ, pfk27Δ, cox5aΔ, and pet191Δ) showed a growth phenotype, but did have a distinguishable metabolic phenotype. This suggests that metabolomics can identify differences between mutants that would be refractory to the common phenotyping methods that are less sensitive. Their other important observation was that the pfk26Δ and pfk27Δ

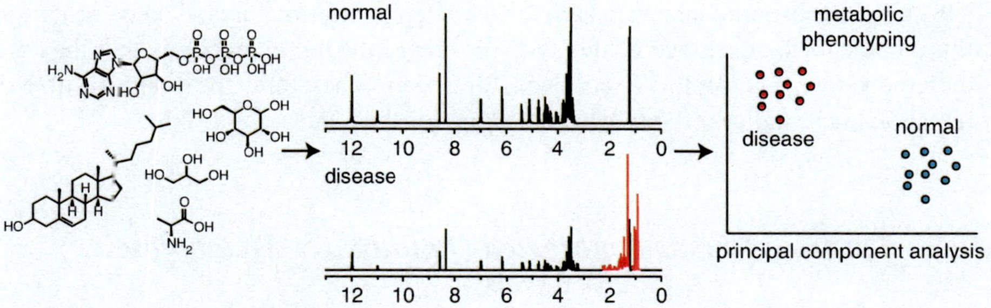

Fig. 2 NMR-based metabolomics can be used to quickly identify changes in the global NMR pattern. In this case, the *red* peaks between 2.5–0.5 ppm are indicative of metabolic differences that are specific to the disease state. Actual data is not nearly as clear as this schematic. The analysis of typical NMR metabolomics datasets requires the use of multivariate analysis methods, such as principle components analysis (PCA), in order to use the metabolome to classify samples

strains, which target the same metabolic pathway, showed similar patterns in their intracellular metabolite levels. Thus, metabolomics comparison of the mutant strains of uncharacterized enzymes and characterized enzymes will enable the function of the uncharacterized proteins to be determined if it is similar to the pattern seen with a known enzyme.

NMR metabolomics has successfully been applied to a number of other systems and biological problems as well. Pioneering work from the Nicholson laboratory, for example, has demonstrated the utility of metabolomics in the investigation of drug metabolism that is influenced by the gut microbiome [32]. By measuring pre-dose levels of urinary metabolites by 1H-NMR, this work was able to find a pattern that would then predict the subsequent metabolism of acetaminophen. Specifically, the presence of high amounts of sulfated *para*-cresol in urine pre-dose meant that the subsequent ratios of sulfated acetaminophen to glucoronidated acetaminophen would be lower after dosage with acetaminophen. Their hypothesis was that the bacterially generated *para*-cresol was acting as a competitive substrate with acetaminophen for sulfating enzymes, which led to the lower ratios of sulfated acetaminophen. This work highlights the use of NMR metabolomics in exploring new problems, such as the metabolic effects of the gut microbiome on the function of drug metabolizing enzymes.

1.4 MS- and LC-MS-Based Metabolomics

Like NMR metabolomics, MS metabolomics can also be used to look broadly across the metabolome in search of patterns that can be used to phenotype organisms. The work of Koulman and colleagues used high resolution Fourier transform mass spectrometry (FTMS) to enable the direct infusion (no chromatography) of fungal metabolites from different strains of endophytic fungi [33]. This work was able to use the metabolite profiles to distinguish different strains of fungi, and tandem

MS (MS/MS) experiments were able to characterize the structures of some of ions in this sample. The advantage of this type of direct-infusion approach is that they are extremely fast, but depth is sacrificed for speed. Therefore, the majority of MS metabolomics studies rely on hyphenated approaches, including LC-MS.

1.5 Targeted Versus Untargeted Quantitative Metabolomics

Typically, metabolomics experiments are performed in either the targeted [27] or untargeted [5] mode depending on what type of information is required (Fig. 3).

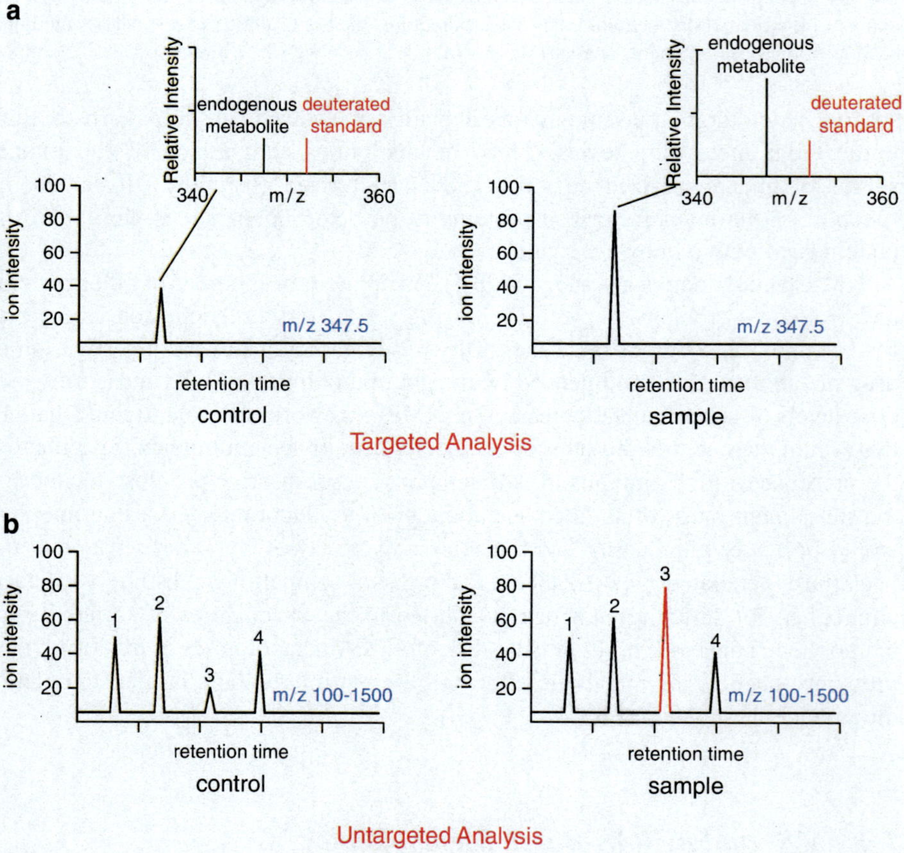

Fig. 3 Comparison of targeted and untargeted metabolomics experiments in the analysis of metabolic differences tied to genotypic changes (i.e., sample versus control). (**a**) Targeted analysis uses isotope dilution mass spectrometry (IDMS), which relies on stable isotope internal standards for quantification of a few metabolites in an absolute state. (**b**) Untargeted analysis measures changes in all metabolites by scanning a broad mass range (e.g., 100–1,500) and quantifying differences based on the ion intensity (e.g., ion *3* increases in the sample). In practice, a combination of both methods is typically the best. For example, we use untargeted metabolomics to scan broadly and discover potential metabolite biomarkers and then follow these experiments up with targeted measurements for absolute quantitation

Targeted analysis refers to metabolome analysis that targets one, or a few metabolites, and typically uses an internal standard for quantitation. The most common method is isotope dilution mass spectrometry (IDMS) [34], which relies on the use of stable isotope internal standards to enable the absolute quantitation of metabolites. This method has proven highly effective and has been successfully used in numerous studies.

IDMS, however, requires that the metabolites that are being measured are known. This limitation prevents the levels of structurally novel metabolites – which give rise to unique ions – to be measured. By contrast, untargeted analysis scans across the entire metabolome and measures the amount of a metabolite based on the ion intensity in the mass spectrometer (i.e., the area under the curve of the peak in the LC-MS) [5, 24]. Untargeted metabolomics contains all of the information found in a targeted experiment, including relative differences between two samples; however, the use of an internal standard during targeted measurements leads to slightly better quantitation.

1.6 Types of Protein–Metabolite Interactions

While the biological sciences have seen an explosion in global methods to identify the molecules associated with a biological process, complementary methods for understanding the mechanisms of a process have been more difficult to develop. Indeed, our own work with activity-based protein profiling (ABPP) has made us keenly aware of the need for a general method to identify interactions between proteins and metabolites [12, 13] (Fig. 4a). One application of metabolomics has been to elucidate protein–metabolite interactions (PMIs), including enzyme–substrate interactions. Of course, biology provides us with a multitude of other important PMIs, such as the binding of metabolites to receptors [35–40] and transport proteins [41]. As a result we have recently extended our metabolomics platform to include these other important, yet hard to identify, PMIs. Together, these examples will highlight the value of metabolomics as a tool for biological discovery.

2 Enzyme–Substrate Interactions

2.1 Classical Approaches and Challenges for Characterizing Enzyme–Substrate Interactions

Substrate characterization typically occurs through in vitro experiments with purified enzymes and model substrates [42–46]. These approaches are often successful at identifying candidate substrates. For example, in the peptidase field, the

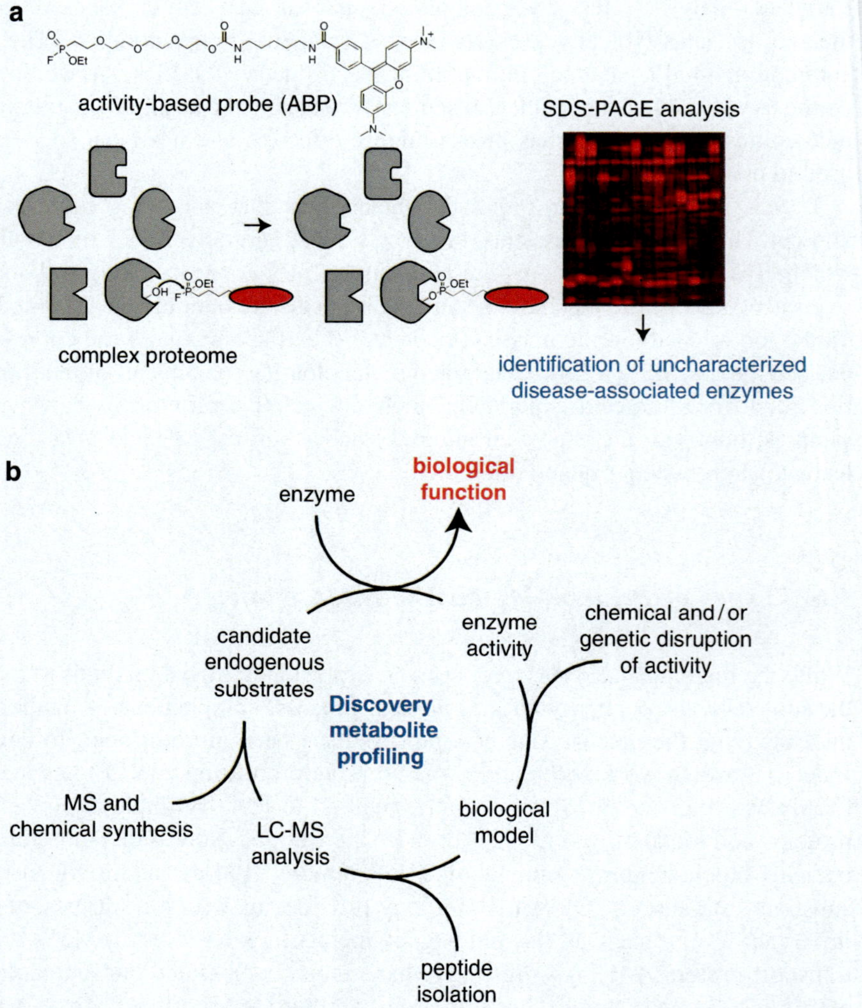

Fig. 4 ABPP and the need for DMP. (**a**) ABPP of the serine hydrolase (SH) superfamily using the fluorophosphonate-rhodamine (FP-Rh) activity-based probe (ABP) to detect active enzymes in complex proteomes. The fluorophosphonate group directs the reactivity of the ABP to the serine hydrolase superfamily, while the rhodamine serves as a visualization tag for analysis of the labeled enzymes by SDS-PAGE. Comparison of samples that vary based on disease state, or other phenotype, can identify SHs with activities that correlate with the phenotype. In some cases these SHs are uncharacterized, which led to the development of DMP as a metabolomics approach to elucidate the natural substrates of enzymes. (**b**) Assigning enzyme function using DMP begins by disruption of the enzyme activity of interest through either chemical or genetic means. Changes in the metabolite profile between disrupted and undisrupted systems are then identified, using LC-MS, and converted into a chemical structure through further MS analysis, FTMS and multiple-stage MS approaches, and chemical synthesis. These compounds are then directly examined as substrates for the enzyme of interest

Table 1 The discordance between in vitro and in vivo substrate selectivity of FAAH

Substrate	k_{cat} (s^{-1})	K_m (μM)	k_{cat}/K_m (M^{-1} s^{-1})	$(k_{cat}/K_m)/(k_{cat}/K_m)$ C18:1 NAE	FAAH($^{-/-}$)/ FAAH($^{+/+}$) brain
C18:1 NAE	8 ± 0.3	35 ± 4	2.3×10^5	1	12.8
C18:1 MAG	3 ± 0.2	28 ± 6	1.1×10^5	0.48	0.72
C18:1 NAT	1.8 ± 0.1	25 ± 5	7.2×10^4	0.31	2.0
C22:0 NAE	≥0.0006	≥150	4.3	2×10^{-5}	≥3.8
C24:0 NAT	≥0.017	≥150	110	5×10^{-4}	26

cleavage specificity of an enzyme is determined using a synthetic peptide substrate library [42, 47]. Once a cleavage preference has been identified, natural peptides are then tested with the peptide to identify candidate endogenous substrates. Often, these in vitro assays fail to identify physiologically relevant enzyme–substrate interactions because they cannot account for important aspects in endogenous biology. These aspects include, but are not limited to, spatial localization of the enzyme that restrict its substrate profile, post-translational modifications of the enzyme or substrate that can change the substrate specificity, competition between enzymes [48], and competition between substrates [49].

As an example of enzyme competition, fatty acid amide hydrolase (FAAH) is able to efficiently hydrolyze monoacylglycerols (MAGs) in vitro; however, in vivo measurements in mice lacking FAAH (FAAH$^{-/-}$ mice) revealed no changes in MAG levels [4]. This lack of an effect is due to the fact that monoacylglycerol lipase (MAGL) is the primary enzyme responsible for the regulation of MAG levels in tissues [50]. Indeed, we found discordance between in vitro substrate specificity and endogenous substrate utilization upon performing metabolomics experiments in FAAH$^{-/-}$ mice (Table 1).

2.2 *Metabolomics Approaches to Characterization of Enzyme–Substrate Interactions*

2.2.1 FAAH

To address the shortcomings associated with traditional methods for characterizing enzyme substrates, metabolomics approaches have recently been applied to this problem. One innovative approach that has been used to identify endogenous substrates is referred to as discovery metabolite profiling (DMP) (Fig. 4b). DMP integrates an array of biological and chemical methods, including genetics, pharmacology, and chemistry (e.g., analytical and synthetic) to quantitatively measure changes in metabolite levels as a function of enzyme activity, to establish connections between enzymes and their endogenous substrates [2–4].

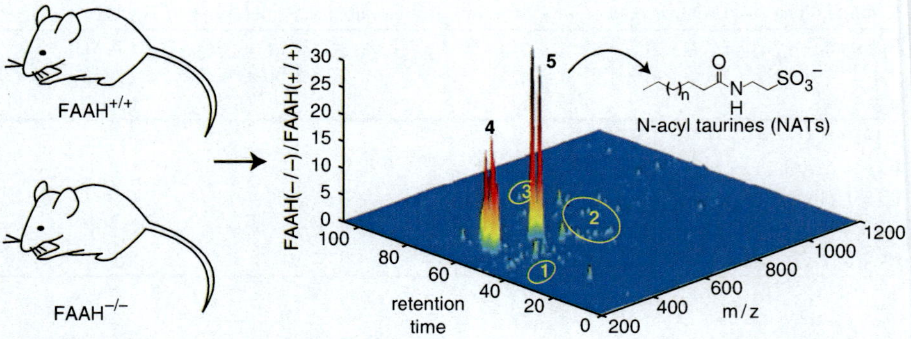

Fig. 5 Discovery metabolite profiling of brain tissue, where mass ion intensity ratios ($FAAH^{-/-}$/ $FAAH^{+/+}$) of metabolites are presented on three-dimensional surface plots. Global view of the relative levels of metabolites in $FAAH^{-/-}$ and $FAAH^{+/+}$ brains, plotted over a mass range of 200–1,200 m/z and liquid chromatography retention times of 0–105 min (plot shown for negative ionization mode). $FAAH^{-/-}$ brains possessed highly elevated levels of *N*-acyl ethanolamines (NAEs) (lipid group *4*) and an unknown class of lipids (group *5*), identified as *N*-acyl taurines (NATs). Other lipids, e.g., free fatty acids (group *1*), phospholipids (group *2*), and ceramides (group *3*) were unaltered in these samples

In contrast to classical, "targeted" LC-MS analysis methods, DMP operates in an untargeted mode, quantifying molecules based on their absolute mass ion intensity (MSII), thereby obviating the need for internal standards and allowing measurements of both known and structurally novel enzyme substrates. We originally developed and applied DMP to the analysis of tissues obtained from mice lacking FAAH [51], an enzyme responsible for degrading the *N*-acyl ethanolamine (NAE) family of signaling lipids (e.g., the endocannabinoid anandamide) [51, 52].

DMP quantified a ~tenfold increase in NAE levels in FAAH-knockout ($FAAH^{-/-}$) mice, as compared to wild type ($FAAH^{+/+}$) mice, an estimate that matched closely with measurements made by targeted methods using isotopic standards [53] (Fig. 5). Remarkably, in addition to detecting known FAAH substrates, DMP also identified a large number of changing ions in these samples that belonged to metabolites of unknown structure. These metabolites were isolated and subsequently characterized through a combination of MS and chemical synthesis [4, 29].

FTMS provided the exact mass of these groups of ions to reveal that they had a variable hydrocarbon region and a conserved heteroatom group (Fig. 6). Tandem MS experiments also revealed a distinctive fragmentation pattern of 124, 107, and 80, which had been seen before for lipids that contained a taurine amide (Fig. 6) [54]. These fragments corresponded to the breakdown of the amide to liberate free taurine (*m*/*z* 124) followed by the breakdown of the taurine to produce the other ions. Based on FAAH's preference for fatty acid amides, this MS data indicated that these novel metabolites were *N*-acyl taurines (NATs). We confirmed this structural assignment through the chemical synthesis of these NATs along with co-elution and tandem MS experiments. This was the first report of the NATs and demonstrated the utility of untargeted metabolomics to discover novel mammalian metabolites, and novel FAAH substrates.

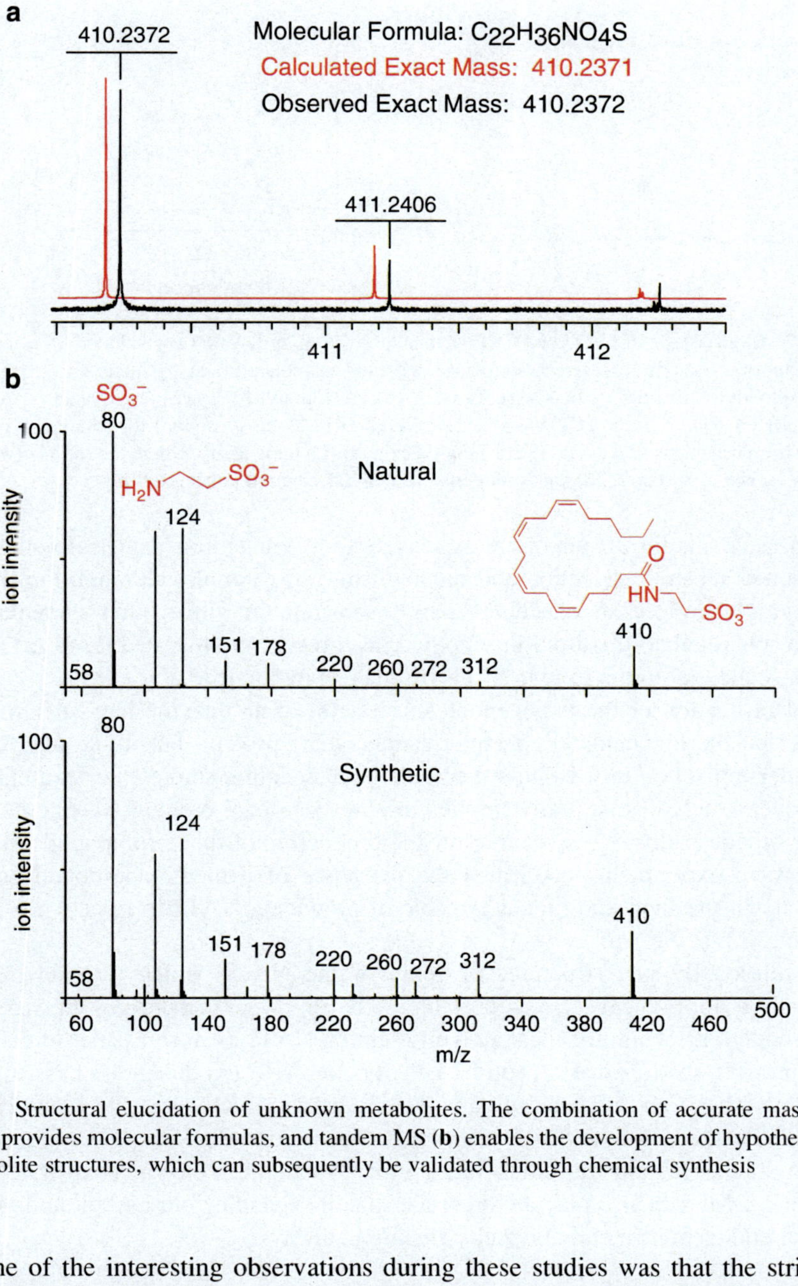

Fig. 6 Structural elucidation of unknown metabolites. The combination of accurate mass (**a**), which provides molecular formulas, and tandem MS (**b**) enables the development of hypotheses of metabolite structures, which can subsequently be validated through chemical synthesis

One of the interesting observations during these studies was that the striking discordance between the in vitro substrate selectivity profile for FAAH and elevated levels of its endogenous substrates in tissue from $FAAH^{-/-}$ mice [5]. This data showed that even very poor in vitro substrates were highly elevated in $FAAH^{-/-}$ mice, while conversely, other lipids that were excellent substrates in vitro (e.g., MAGs) were unperturbed by the inactivation of FAAH in vivo (Table 1). These differences can be rationalized by realizing that, in contrast to in vitro methods,

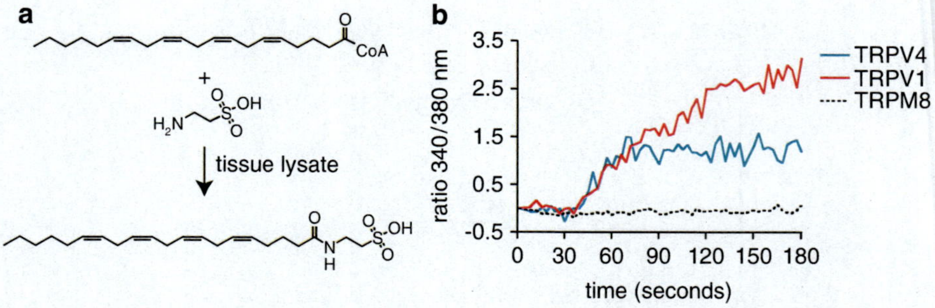

Fig. 7 Biosynthesis of NATs and TRP channel activation by NATs. (**a**) Evidence for a fatty acyl CoA:taurine *N*-acyltransferase activity was detected in mouse tissue by incubating taurine and arachidonoyl-CoA with various tissue lysates. (**b**) arachidonyl NAT was tested as an activator of the TRPV1 (*black line*), TRPV4 (*gray line*), and TRPM8 (*dashed line*) ion channels. Channel activation was measured using a Fura-2-based calcium-imaging assay, where the ratio between the fluorescence at 340 and 380 nm is reflective of cellular calcium concentrations

DMP accounts for *all* important aspects of endogenous biochemistry/biology that influence substrate selection and metabolism. For example, although the lipid 2-oleoyl glycerol is an excellent FAAH substrate in vitro, other enzymes and pathways regulate this lipid in vivo and, as a result, the loss of FAAH activity is inconsequential to the in vivo levels of this metabolite [5].

Identification of the endogenous substrate(s) of an enzyme has an immediate effect on the biochemical characterization of the protein, but it can also have a broader impact on downstream metabolic and signaling studies. For example, the discovery of NATs naturally implies the presence of a previously unappreciated biosynthetic pathway responsible for the production of these compounds. Indeed, follow-up experiments confirmed the existence of a membrane-bound enzyme activity in the liver and kidney capable of producing NATs from acyl-CoAs and taurines [29] (Fig. 7a).

Additionally, the structures of some of the NATs, which resemble known signaling lipids, suggested a possible role for these metabolites in signaling. Subsequent screening of these polyunsaturated NATs against a panel of different receptor classes (nuclear receptors, GPCRs, and TRP ion channels) has validated this hypothesis and revealed that the NATs activate TRPV ion channels (TRPV1 and TRPV4) at physiologically relevant concentrations (Fig. 7b) [29]. In aggregate, these studies demonstrate that, in addition to assigning enzyme–substrate interactions, DMP can also play an important role in expanding our current understanding of endogenous metabolic and signaling pathways.

2.2.2 KIAA

DMP was primarily developed to enable the analysis of unknown, or uncharacterized, enzymes discovered through ABPP. Jessani, Cravatt, and colleagues

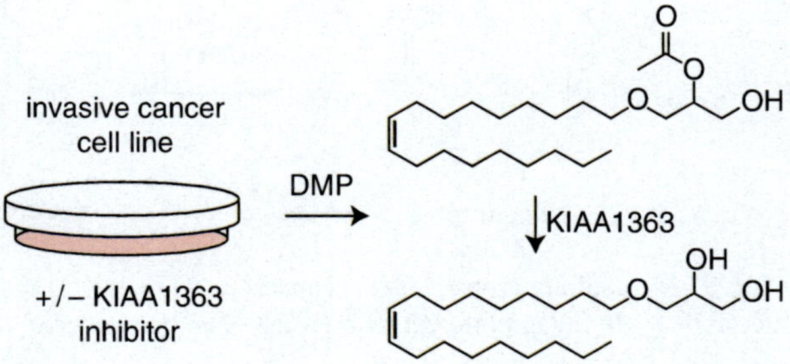

Fig. 8 An untargeted LC-MS-based metabolomics approach revealed that KIAA1363, an uncharacterized invasiveness-related hydrolase, regulates MAGE levels in cancer cell lines

had used ABPP to profile the enzyme activities in a series of cancer cell lines [12]. These cells differed in their ability to invade a matrigel matrix in vitro, which is thought to correlate with invasiveness and metastatic ability in vivo. Interestingly, these experiments revealed a novel enzyme KIAA1363, which was strongly active in invasive cell lines from many different tissues of origin. Homology indicated that KIAA1363 was a lipase, but none of the closely related KIAA1363 homologs had been characterized [12]. As a result, DMP was used in some of these cell lines to identify the substrates and pathways regulated by KIAA1363 [24].

Metabolomic comparison of untreated cells with cells treated with the KIAA1363 inhibitor, AS115, identified a number of changes in the AS115-treated metabolomes. In this case, inhibitor treatment led to lower levels of a number of metabolites, and these metabolites were subsequently identified as monoaklyl glycerol ethers (MAGEs). Since these MAGEs had lower levels in the absence of KIAA1363 activity they were identified as products of the enzyme. Analysis of candidate substrates revealed that KIAA1363 removes the acetyl group from the two position of 2-actetyl MAGE to produce MAGE (Fig. 8).

Work by Snyder and colleagues had identified MAGEs as being upregulated in a number of cancers [55]. Further evidence for this connection between KIAA1363 and MAGE was obtained when it was shown that cells that had elevated MAGE levels also had elevated KIAA1363 levels. By identifying a cellularly relevant enzyme–substrate interaction, DMP enabled the impact of MAGE levels in cancer to be assessed by looking at cells with different KIAA activities. Inhibition of KIAA led to lower levels of MAGE and reduced invasiveness in an in vitro matrigel assay as well as in an in vivo xenograft study. This data demonstrates the value of metabolomics in the identification of enzyme–substrate interactions.

2.2.3 IDH

In addition to these lipophilic metabolomics experiments, similar approaches can be used to measure differences in hydrophilic metabolites. One interesting example of this comes from the desire to characterize a mutation found in isocitrate dehydrogenase (IDH) in a number of different cancers [31, 56, 57]. In this case, the switch of arginine 132 to histidine (R132H) was found in ~80% of all grade II and III gliomas and glioblastomas. Interestingly, a mutation in even a single copy of this

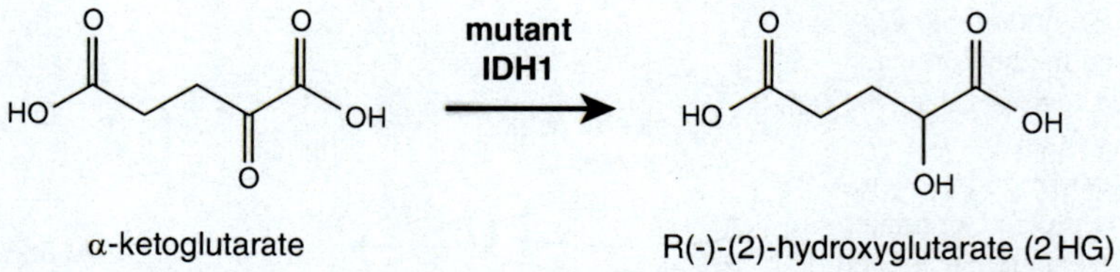

Fig. 9 Metabolomics revealed that cancer-associated IDH1 mutations enable the enzyme to begin to produce *R*(-)-(2)-hydroxyglutarate (2HG), an oncometabolite

gene led to cancer, suggesting that this was a gain-of-function mutation. Metabolomics of U87MG cells transfected with wild-type or mutant (R132H) IDH was used to identify what changes, if any, in metabolism was being caused by IDH.

This analysis revealed a set of ions that were strongly elevated in the IDH mutant-containing cell lines. The exact mass identified these ions as (*R*)-2-hydroxyglutarate (2HG), the sodium adduct of 2-HG, and the dehydrated product of 2-HG. In vitro experiments with recombinant IDH confirmed that the mutant enzyme, but not the wild-type enzyme, is able to produce 2-HG (Fig. 9). Finally, this correlation was demonstrated to be of clinical relevance because patients with mutated IDH had higher levels of 2-HG. This example highlights the utility of metabolomics in the characterization of enzyme–substrate interactions, in this case a mutated enzyme associated with cancer. Moreover, this work has opened up a new field dedicated towards understanding if and how this IDH mutation leads to cancer [31, 56, 57].

3 Other Protein–Metabolite Interactions

3.1 Significance

Biology uses a multitude of different molecular interactions (e.g., protein–protein [58], protein–DNA [59], etc.) to control cellular and physiological processes. Amongst these interactions, connections between proteins and small molecule metabolites are some of the most important. Indeed, PMIs are at the center of processes as diverse as development (steroids) [60], inflammation (prostaglandins) [61], and pain (endocannabinoids) [62]. In many of these situations, the metabolite is often used as a diffusible signal (i.e., a hormone) that regulates cellular and physiological phenotypes upon binding to a protein receptor. In addition, other proteins have also evolved to help facilitate the function of metabolites by transporting metabolites within cells [41], through cellular membranes [63], or between tissues [64]. For example, certain fatty acid binding proteins are responsible for shuttling cytosolic fatty acids to the cell nucleus where these metabolites regulate gene transcription [65, 66].

Importantly, in many instances, understanding which PMIs regulate a biological process can provide insights necessary for the development of novel therapeutics, because many of these proteins are very druggable, owing to their natural ability to bind small molecules. For example, the discovery that arachidonyl ethanolamide (anandamide) is the endogenous ligand for the cannabinoid receptor has led to a serious effort in the pharmaceutical industry to develop compounds that raise anandamide levels [67] for the treatment of chronic pain. Despite the importance of PMIs in almost every aspect of biology and biomedicine, there have been relatively few methods developed for studying these interactions [68]. As a result, we anticipate that the development of modern approaches to elucidate PMIs will have an immediate impact on our ability to discover PMIs, characterize the proteins involved in these interactions, and eventually lead to the development of new medicines.

3.2 Classical Methods for PMIs

Numerous approaches have proven useful in the discovery of PMIs. The development of binding assays for the discovery of PMIs often depends on the function of the protein. In one example, ligands for the START domain 3 (StarD3) protein, a putative lipid binding protein, was identified using various biophysical techniques including binding of radiolabeled sterols [69]. StarD3 was immobilized onto a solid support and four different radiolabeled sterols were separately incubated with the bound protein. Measurement of the residual radioactivity after removal of the sterol solution and washing of bound StarD3 identified 25-hydroxycholesterol as a clear binder, while 27-hydroxycholesterol did not bind. While effective, this approach is not general and is limited by the number of radiolabeled metabolites available.

For other small molecule binding proteins, like transcription factors, the read-out of a screen can be coupled to a cell-based assay looking at protein function. For example, the discovery of a candidate ligand for the NOR1 nuclear receptor was accomplished by screening a lipid library using an assay that coupled NOR1 activity to luciferase expression [70]. This approach successfully identified arachidonic acid as an NOR1 agonist. Similar types of screens have successfully been used to identify natural and synthetic ligands for additional nuclear receptors, such as Nur77 [71] and FXR [72, 73]. While effective, these approaches require ligands that are cell permeable, and cannot be generalized to proteins that do not act as transcription factors.

Another successful approach for the discovery of novel PMIs and signaling molecules has been to use a biochemical screening approach, where the metabolome is fractionated and tested for binding to a protein. For example, the discovery of anandamide as an endogenous ligand for the cannabinoid receptor came from the fractionation and screening of the brain lipids for compounds that could displace radiolabeled tetrahydrocannabinol (THC) from the cannabinoid receptor [62]. Of course, as with any approach, screening-based approaches to PMI discovery also have some limitations. Lastly, screens are only designed to find a molecule with a particular property, such as an activator of a receptor. As a consequence, these screens often miss protein binders that might inhibit a particular protein. Thus, there is room for the development of new approaches for the discovery of PMIs that can overcome some of these challenges.

3.3 *Metabolomics-Based Approach*

DMP highlights the power of metabolomics in assigning enzyme–substrate interactions. While the DMP approach is specific to enzyme–substrate interactions, we felt that the untargeted metabolomics platform at the center of DMP could be integrated into another workflow to enable the discovery of non-enzymatic PMIs. We reasoned that if we could develop an efficient strategy for the isolation of a protein–metabolite complex [69], the metabolomics platform could then be used to identify any protein bound ligands (Fig. 10). Initially, we imagined a method where a "pull-down" step would be used to isolate a protein metabolite complex from a cell or tissue lysate, followed by metabolomics. This approach would be directly analogous to proteomics-based methods for identifying protein–protein interactions where a particular protein is immunoprecipitated and the bound proteins are identified by proteomics [74].

In theory, isolation and direct analysis of a protein metabolite complex would afford a general approach for identifying PMIs, and could overcome many of the challenges associated with screening-based approaches, including the necessity for radiolabeled metabolites for particular assays. Prior to attempting the isolation of protein–metabolite complexes directly from cells and tissues, however, we realized that there were some major technical issues that we needed to resolve first. One concern was that PMIs are typically much weaker interactions than protein–protein interactions and, therefore, would be more likely to dissociate during the isolation of a protein–metabolite complex. Additionally, attempting to immunoprecipitate proteins directly from lysates would introduce an extra step that would complicate the development of a general metabolomics PMI approach.

The primary steps in this approach begin with the recombinant expression of the protein with a tag that will allow us to immobilize the protein. To date, we've used glutathione *S*-transferase (GST) or polyhistidine (His6) affinity handles to immobilize proteins onto a solid support (Fig. 10). The bound protein is subsequently incubated with a mixture of cellular metabolites that are isolated from cells and tissues where the protein is naturally expressed. During the incubation step, metabolites can bind to the protein resulting in a protein–metabolite complex on the solid support. In this step, we are essentially screening the thousands of metabolites in parallel to identify a binder, which greatly accelerates the rate over fractionation-based methods and should improve the chances of discovering reactive metabolites.

The immobilized protein metabolite complex is then filtered to remove unbound metabolites and the beads are quickly washed (5–10 s) to minimize the loss of

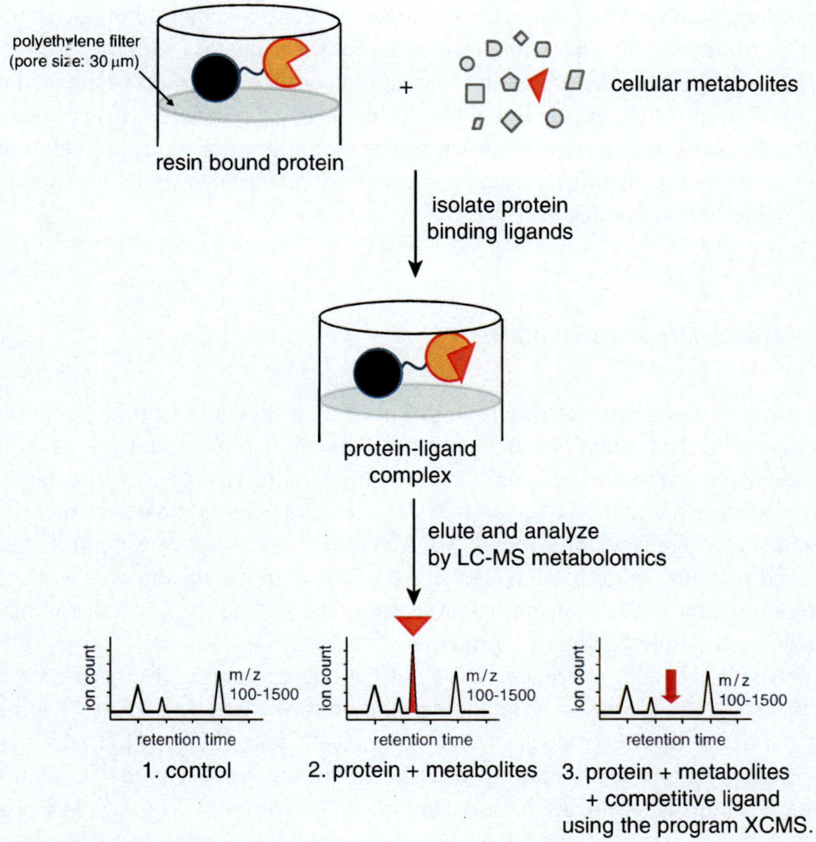

Fig. 10 A liquid chromatography-mass spectrometry (LC-MS)-based metabolomics approach for characterizing endogenous nuclear receptor (NR) ligands. *Top*: In the first step, metabolites isolated from cells or tissues are incubated with a ligand-binding domain (LBD) of an NR that has been immobilized onto a solid support. *Middle*: During the incubation, any endogenous ligands can bind to the NR and form a protein–metabolite complex. *Bottom*: Unbound metabolites are then washed away, and the NR–ligand complex is eluted from the resin. NR-bound metabolites are identified by comparing the LC-MS chromatograms from NR samples to control samples (no protein, competitive ligand using the program XCMS

bound metabolites. Next, the protein–metabolite complex is eluted from the beads and this eluate is then directly analyzed using LC-MS metabolomics, which generates a list of ions corresponding to metabolites present on the beads. To identify ions specifically enriched by the metabolite binding protein, our data analysis step compares GST–protein to GST alone, which identifies ions specifically retained by the protein. In addition, we have found that an additional comparison between GST–protein and GST–protein (no lipid added) helps account for lipids that co-purify with the recombinant protein.

Finally, we will identify the structure of enriched ions by using accurate mass (obtained during our standard MS experiments) and searching metabolite databases, such as METLIN [75], PubChem [76], or the lipid MAPS [77] for metabolites with similar molecular formulas. In many instances though we will be required to structurally characterize enriched ions. As detailed with the discovery of the NATs above, the identification of novel metabolites can be accomplished using a combination of MS and chemical synthesis [4, 78]. To develop this approach we decided to focus on a group of known lipid binding proteins (LBPs), which would allow us to focus our efforts directly on the key factors necessary for identifying PMIs using metabolomics.

3.4 Lipid-Binding Proteins

The initial experiments began with a group of three LBPs to test the generality of the approach, and identify any potential bias in the methodology. The LBPs selected were cytosolic retinoic acid binding protein (CRABP2) [79, 80], fatty acid binding protein (FABP), and StarD3, a cholesterol binder. Expression of CRABP2 as a GST fusion protein (CRABP-GST) was readily accomplished and provided us with active protein. For these LPBs, activity was measured using the hydrophobic dye 1-anilinonaphthalene-8-sulfonate (ANS) [81], which can be used to measure the folded state of a protein.

CRABP-GST is immobilized onto a glutathione-containing resin at a concentration that saturates the resin. A "complex lipid mixture" was prepared by combining a concentrated lipid extract derived from mouse tissue (brain)[5] (DMSO stock) with exogenously added retinoic acid (RA, positive control) and ^{13}C-oleic acid (negative control). Analysis of this "complex lipid mixture" by LC-MS prior to incubation with immobilized CRABP demonstrated that the mixture contained RA (added), phospholipids, acylglycerols, cholesterol esters, and cholesterol. A portion of this mixture (corresponding to 1 nmol of RA) was added to PBS (DMSO concentration should not exceed 5%) and incubated with the CRABP-GST bound beads for 1 h.

After incubation, the beads were washed, and the CRABP-GST was eluted through the addition of glutathione-containing buffer. We found that elution of the protein, rather than direct extraction of the ligand from the beads, led to lower background levels due to organic contaminants that are infused into the beads. The protein-containing sample and the two control samples (no lipid and no protein) were analyzed using LC-MS metabolomics. These samples were then processed with XCMS [82], which aligns, quantifies, and statistically ranks differences between LC-MS data sets, to show differences in metabolite levels between the CRABP-GST sample and the controls (no protein, no lipid) to identify CRABP binding partners.

After analysis of the CRABP-GST samples with XCMS, two statistically significant ions (*m/z* 299 and 279) were identified. Based on co-elution with known standards and accurate mass, these ions were confirmed to correspond to RA and linoleic acid (C18:2 free fatty acid) (Fig. 11). The fold change for RA (~14-fold) is much larger than for linoleic acid (twofold), as would be expected for CRABP.

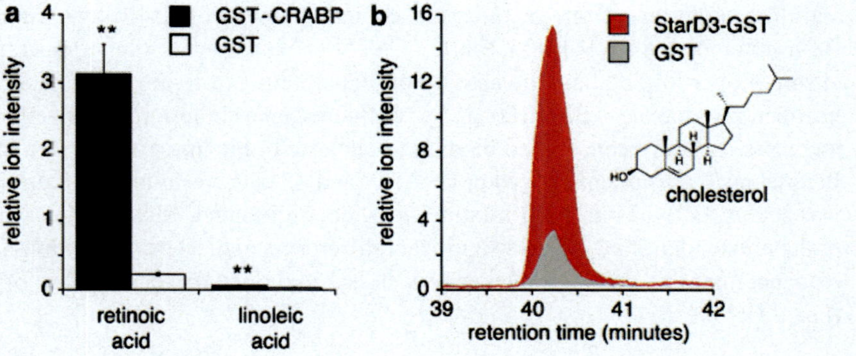

Fig. 11 Data from metabolomics PMI experiments. (**a**) Enrichment of retinoic acid by CRABP-GST from a mixture of brain lipids (**$P < 0.01$, $n = 3$–4, Student's t-test). (**b**) StarD3-mediated enrichment of cholesterol from a brain lipid extract

Table 2 Identification of specific binding ligands of FABP2 from brain lipid extracts

Lipid	FABP-GST (avg. ion counts)	GST (avg. ion counts)	Fold change
C18:2 FFA	2.72×10^4	4.32×10^3	6.3*
^{12}C-C18:1 FFA	3.24×10^5	8.27×10^3	39.2*
^{13}C-C18:1 FFA	9.10×10^5	5.90×10^4	15.4*
C20:4 FFA	2.46×10^5	7.32×10^4	3.4*
Retinoic acid	1.45×10^5	1.17×10^5	1.2

*$P < 0.05$, $n = 3$–4, Student's t-test

These results validate our data analysis method and, more generally, highlight the utility of metabolomics to identify natural PMIs from complex lipid mixtures.

The next goal was to determine the generality of this approach by testing other classes of LBPs. FABP and StarD3 were chosen because their ligands are very different from each other, and different from RA. FABP is an intestinal protein that transports fatty acids within cells and has been linked to metabolic disorders and cardiovascular disease [83, 84]. FABP2 was expressed as a GST fusion protein (FABP-GST) and the activity of this protein was confirmed by measuring oleic acid, a known ligand by displacement of ANS.

Resin-bound FABP-GST was incubated with the "complex lipid mixture." After incubation for 1 h, the mixture was removed, the beads washed, and the protein eluted through addition of glutathione. Analysis of the eluate by global metabolite profiling and XCMS revealed the specific enrichment of fatty acids by FABP from the lipid extracts (Table 2). More specifically, oleic acid, linoleic acid, and arachidonic acid were all enriched and this is consistent with observed lipid specificity of FABP2. This example demonstrates the ability of PMI metabolomics to identify natural binding partners from complex mixtures of natural lipids.

StarD3 was examined to ensure that the methodology is not biased towards anionic carboxylate-containing lipids. StarD3 is a member of the StarD family of LBPs that are found in mammalian tissues and are responsible for binding and shuttling cholesterol from the outer plasma membrane to organelles within the cell for use in metabolism [41, 85]. StarD3-GST was expressed at room temperature to ensure proper folding, and the activity of this protein was tested using cholesterol-nitrobenzoxadiazole (ch-NBD) [85], a fluorescent cholesterol derivative that increases its fluorescence upon binding to cholesterol binding proteins like StarD3. In analogy to experiments with CRABP and FABP, resin-bound StarD3-GST was incubated with the lipid mixture, washed, and eluted. Metabolomic analysis of the eluate identified a single significant difference, cholesterol, providing further evidence that this metabolomics approach can assign PMIs for a variety of LBPs (Fig. 11).

3.5 Nuclear Receptors PPARγ and PPARα

Nuclear receptors are an extremely interesting class of transcription factors that control a wide range of physiology [60]. The protein structure of nuclear receptors can be divided into two primary domains: a DNA-binding domain (DBD) and a ligand-binding domain (LBD). Binding of small molecules, primarily lipids, to the LBD often leads to the increased transcriptional activity of the nuclear receptor. The molecules that bind nuclear receptors include vitamins (e.g., vitamin D receptor [86]), hormones (e.g., estrogen receptor [87]), and approved drugs, such as the PPARγ agonist rosiglitazone [88] (Avandia). In humans there are 48 nuclear receptors [89] and, while many of these have a known ligand [60, 90], a number of receptors are referred to as orphan nuclear receptors because they lack any known natural ligand [90] (some of these orphan receptors do have synthetic ligands).

The identification of ligands for orphan receptors is an active research area with important implications in understanding molecular pathways related to physiology and medicine [91, 92]. New natural and synthetic ligands for nuclear receptors are typically found by screening small molecules using a reporter gene assay (luciferase production). For example, the peroxisome proliferator-activated receptor γ (PPARγ) ligand 15-deoxy-Δ-12,14-prostaglandin-J2 (15d-PGJ2) was identified [35] by searching for arachidonic acid metabolites that activate PPAR response elements. The identification of the candidate nuclear receptor ligands via this approach can be challenging, especially if a robust assay is not available, or the candidate ligands are not known.

The studies with LBPs enabled us to develop the workflow necessary for elucidating PMIs. To analyze nuclear receptors using the same approach we faced a challenge in that LBPs tend to bind their ligands more tightly (typical K_ds are in the nanomolar range) [81, 83] than nuclear receptors (some K_ds are in micromolar range) [35, 93]. PPARγ and PPARα were analyzed to see whether PMI metabolomics can be extended to the discovery of natural ligands for nuclear receptors. PPARγ is an ideal choice because there are a number of known ligands [93] that can serve as positive controls. In addition, PPARγ is a target of the insulin-sensitizing thiazolidinedione class of drugs [94], and these synthetic ligands can be used to as controls to validate that the PPARγ is in an active conformation. Since the ligand-binding domains (LBD) of nuclear receptors are sufficient for ligand binding, we specifically cloned and expressed the LBDs.

A purified polyhistidine-tagged PPARγ-LBD (His6-PPARγ-LBD) was immobilized onto a metal affinity chromatography (IMAC) resin. The immobilized His6-PPARγ-LBD was incubated with brain or liver lipid extracts. After incubation, the beads were washed and the protein was eluted from the solid support. Metabolomics analysis and comparison of the His6-PPARγ-LBD sample to the no-protein control (i.e., beads only) using XCMS identified a number of PPARγ ligands (Table 3). Specifically, arachidonic acid was enriched from the brain lipid extract and oleoyl lysophosphatidic acid (LPA) was enriched from the brain and liver lipid extracts. Higher concentrations of polyunsaturated fatty acids in the nervous system compared with the liver can explain the absence of arachidonic acid in the liver samples [29].

PPARγ is also known to play a role in adipose tissue formation and the ligands might differ in a tissue-dependent manner. To account for this, adipose tissue extracts were also examined using PMI metabolomics. Immobilized His6-PPARγ-LBD was incubated with this fractionated adipose tissue extract, which led to the enrichment of a number of fatty acids, including arachidonic, linoleic, and oleic acids (Table 3). Similar results were obtained in the analysis of 3T3-L1, and adipose cell line extracts. These lipids have all been reported as PPARγ ligands [93], demonstrating the ability of our methodology to identify bona fide ligands of a nuclear receptor from tissues.

The generality of this approach was tested by looking for lipids enriched by PPARα, another member of the PPAR family of nuclear receptors. His6-PPARα-LBD

Table 3 PPARγ/α ligands characterized by untargeted metabolomics from various tissues and cells

Source	Lipid	Formula	PPARγ/control
Brain	C20:4 FFA	$C_{20}H_{32}O_2$	8.6*
	C18:1 LPA	$C_{21}H_{41}O_7P$	4.4*
Liver	C18:1 LPA	$C_{21}H_{41}O_7P$	6.8*
Fat	C20:4 FFA	$C_{20}H_{32}O_2$	18.1**
	C18:1 FFA	$C_{18}H_{34}O2$	6.3**
	C18:2 FFA	$C_{18}H_{32}O_2$	16.0**
3T3-L1 cell (Day2, DMI)	C20:4 FFA	$C_{20}H_{32}O_2$	5.1*
	C18:1 FFA	$C_{18}H_{32}O_2$	2.4*
	C18:2 FFA	$C_{18}H_{34}O_2$	3.4**
	C16:1 FFA	$C_{16}H_{30}O_2$	2.2*
Source	**Lipid**	**Formula**	**PPARα/control**
Brain	C20:4 FFA	$C_{20}H_{32}O_2$	12.4**
	C18:1 FFA	$C_{18}H_{34}O_2$	2.2*
	C18:2 FFA	$C_{18}H_{32}O_2$	7.1**

$*P < 0.05$, $**P < 0.01$; $n = 3$–4, Student's *t*-test

was expressed, purified, and tested for activity using an ANS binding assay. Since PPARα is expressed in the liver, a lipid extract of the liver was used to identify ligands for PPARα. After incubation of liver extract with His6-PPARα-LBD, the resin was washed, and the protein eluted by the addition of imidazole. Comparison of the protein sample to the no protein control using metabolomics and XCMS revealed that the His6-PPARα-LBD preferentially bound fatty acids, including arachidonic (C20:4), linoleic (C18:2), and oleic (C18:1) acids (Table 3). These are all reported PPARα ligands [93] and this result provides a second example of the ability of this metabolomics strategy to characterize ligands for nuclear receptors from complex lipid mixtures.

3.6 Candidate Ligands and Confirmation

It is important to realize that the identification of the PMI using metabolomics is the first step in the characterization of a ligand. First, it is imperative that secondary biophysical assays be performed to confirm the binding between a ligand and a protein. For example, isothermal calorimetry has been with tremendous success to measure binding of a ligand to a protein. Likewise, other biophysical methods such as changes in the circular dichroism spectrum, shifts in protein thermodynamics in the presence of ligand can be used to establish binding [69]. Other methods can also be employed as long as other assays are used to confirm the biochemical interaction. Additionally, structural characterization using NMR or X-ray crystallography can be used to confirm and study the interaction in greater detail [95].

In addition to these biophysical methods, it is important to test the physiological relevance of these approaches by measuring changes in protein function in the presence or absence of the ligand. Depending on the protein, these assays will clearly vary but must be performed as a means to confirm biological relevance. Finally, the connection between the protein and the ligand must be assessed under cellular and/or physiological conditions where the natural concentration of the ligand is varied. For example, anandamide was confirmed as a ligand for the cannabinoid receptor when changes in its endogenous concentrations led to physiological differences that were dependent on the CB1 receptor [51].

4 Conclusions and Future Directions

Metabolomics has proven to be a useful and efficient approach for the discovery of enzyme–substrate interactions, metabolites, and protein–metabolite interactions. As a result, these metabolomics approaches have proven useful functional proteomics tools, especially when coupled to methods like ABPP. The continued use of these methods should result in improved methodology and validation of the efficiency of these approaches.

Acknowledgement We thank the Searle Scholar Award, Burroughs Wellcome Fund Career Award in the Biomedical Sciences, and National Institutes of Health Grants 1DP2OD002374 for support.

References

1. Berg JM, Tymoczko JL, Stryer L (2007) Biochemistry. W.H. Freeman, New York
2. Saghatelian A, Cravatt BF (2005) Curr Opin Chem Biol 9:62
3. Saghatelian A, Cravatt BF (2005) Life Sci 77:1759
4. Saghatelian A, Cravatt BF (2005) Nat Chem Biol 1:130
5. Saghatelian A, Trauger SA, Want EJ, Hawkins EG, Siuzdak G, Cravatt BF (2004) Biochemistry 43:14332
6. Schena M, Shalon D, Davis RW, Brown PO (1995) Science 270:467
7. Mortazavi A, Williams BA, McCue K, Schaeffer L, Wold B (2008) Nat Methods 5:621
8. Wang Z, Gerstein M, Snyder M (2009) Nat Rev Genet 10:57
9. Delahunty CM, Yates JR III (2007) BioTechniques 43:563
10. Motoyama A, Venable JD, Ruse CI, Yates JR III (2006) Anal Chem 78:5109
11. Washburn MP, Wolters D, Yates JR III (2001) Nat Biotechnol 19:242
12. Jessani N, Liu Y, Humphrey M, Cravatt BF (2002) Proc Natl Acad Sci USA 99:10335
13. Jessani N, Niessen S, Wei BQ, Nicolau M, Humphrey M, Ji Y, Han W, Noh DY, Yates JR III, Jeffrey SS, Cravatt BF (2005) Nat Methods 2:691
14. Raamsdonk LM, Teusink B, Broadhurst D, Zhang N, Hayes A, Walsh MC, Berden JA, Brindle KM, Kell DB, Rowland JJ, Westerhoff HV, van Dam K, Oliver SG (2001) Nat Biotechnol 19:45
15. Fiehn O (2002) Plant Mol Biol 48:155
16. Watkins SM, Reifsnyder PR, Pan HJ, German JB, Leiter EH (2002) J Lipid Res 43:1809
17. Watkins SM, German JB (2002) Curr Opin Biotechnol 13:512
18. Rabinowitz JD (2007) Expert Rev Proteomics 4:187
19. Chadeau-Hyam M, Ebbels TM, Brown IJ, Chan Q, Stamler J, Huang CC, Daviglus ML, Ueshima H, Zhao L, Holmes E, Nicholson JK, Elliott P, De Iorio M (2010) J Proteome Res 9:4620
20. Garrod A (1908) Lancet 172:1
21. Garrod AE (1899) Med Chir Trans 82:367
22. Granadino B, Beltran-Valero de Bernabe D, Fernandez-Canon JM, Penalva MA, Rodriguez de Cordoba S (1997) Genomics 43:115
23. Sreekumar A, Poisson LM, Rajendiran TM, Khan AP, Cao Q, Yu J, Laxman B, Mehra R, Lonigro RJ, Li Y, Nyati MK, Ahsan A, Kalyana-Sundaram S, Han B, Cao X, Byun J, Omenn GS, Ghosh D, Pennathur S, Alexander DC, Berger A, Shuster JR, Wei JT, Varambally S, Beecher C, Chinnaiyan AM (2009) Nature 457:910
24. Chiang KP, Niessen S, Saghatelian A, Cravatt BF (2006) Chem Biol 13:1041
25. Wishart DS, Knox C, Guo AC, Eisner R, Young N, Gautam B, Hau DD, Psychogios N, Dong E, Bouatra S, Mandal R, Sinelnikov I, Xia J, Jia L, Cruz JA, Lim E, Sobsey CA, Shrivastava S, Huang P, Liu P, Fang L, Peng J, Fradette R, Cheng D, Tzur D, Clements M, Lewis A, De Souza A, Zuniga A, Dawe M, Xiong Y, Clive D, Greiner R, Nazyrova A, Shaykhutdinov R, Li L, Vogel HJ, Forsythe I (2009) Nucleic Acids Res 37:D603
26. Dennis EA (2009) Proc Natl Acad Sci USA 106:2089
27. Bennett BD, Yuan J, Kimball EH, Rabinowitz JD (2008) Nat Protoc 3:1299
28. Denkert C, Budczies J, Weichert W, Wohlgemuth G, Scholz M, Kind T, Niesporek S, Noske A, Buckendahl A, Dietel M, Fiehn O (2008) Mol Cancer 7:72

29. Saghatelian A, McKinney MK, Bandell M, Patapoutian A, Cravatt BF (2006) Biochemistry 45:9007
30. Pesek JJ, Matyska MT, Fischer SM, Sana TR (2008) J Chromatogr A 1204:48
31. Ward PS, Patel J, Wise DR, Abdel-Wahab O, Bennett BD, Coller HA, Cross JR, Fantin VR, Hedvat CV, Perl AE, Rabinowitz JD, Carroll M, Su SM, Sharp KA, Levine RL, Thompson CB (2010) Cancer cell 17:225
32. Clayton TA, Baker D, Lindon JC, Everett JR, Nicholson JK (2009) Proc Natl Acad Sci USA 106:14728
33. Koulman A, Tapper BA, Fraser K, Cao M, Lane GA, Rasmussen S (2007) Rapid Commun Mass Spectrom 21:421
34. Dzeletovic S, Breuer O, Lund E, Diczfalusy U (1995) Anal Biochem 225:73
35. Forman BM, Tontonoz P, Chen J, Brun RP, Spiegelman BM, Evans RM (1995) Cell 83:803
36. Vassilatis DK, Hohmann JG, Zeng H, Li F, Ranchalis JE, Mortrud MT, Brown A, Rodriguez SS, Weller JR, Wright AC, Bergmann JE, Gaitanaris GA (2003) Proc Natl Acad Sci USA 100:4903
37. Peier AM, Moqrich A, Hergarden AC, Reeve AJ, Andersson DA, Story GM, Earley TJ, Dragoni I, McIntyre P, Bevan S, Patapoutian A (2002) Cell 108:705
38. Caterina MJ, Schumacher MA, Tominaga M, Rosen TA, Levine JD, Julius D (1997) Nature 389:816
39. Ortlund EA, Lee Y, Solomon IH, Hager JM, Safi R, Choi Y, Guan Z, Tripathy A, Raetz CR, McDonnell DP, Moore DD, Redinbo MR (2005) Nat Struct Mol Biol 12:357
40. Rosen H, Liao J (2003) Curr Opin Chem Biol 7:461
41. Alpy F, Tomasetto C (2005) J Cell Sci 118:2791
42. Backes BJ, Harris JL, Leonetti F, Craik CS, Ellman JA (2000) Nat Biotechnol 18:187
43. Jenkins CM, Mancuso DJ, Yan W, Sims HF, Gibson B, Gross RW (2004) J Biol Chem 279:48968
44. Flint AJ, Tiganis T, Barford D, Tonks NK (1997) Proc Natl Acad Sci USA 94:1680
45. Favia AD, Nobeli I, Glaser F, Thornton JM (2008) J Mol Biol 375:855
46. Kashiwayama Y, Tomohiro T, Narita K, Suzumura M, Glumoff T, Hiltunen JK, Van Veldhoven PP, Hatanaka Y, Imanaka T (2010) J Biol Chem 285:26315
47. Leiting B, Pryor KD, Wu JK, Marsilio F, Patel RA, Craik CS, Ellman JA, Cummings RT, Thornberry NA (2003) Biochem J 371:525
48. Bishop AC, Shah K, Liu Y, Witucki L, Kung C, Shokat KM (1998) Curr Biol 8:257
49. Tinoco AD, Kim YG, Tagore DM, Wiwczar J, Lane WS, Danial NN, Saghatelian A (2011) Biochemistry 50:2213
50. Long JZ, Nomura DK, Cravatt BF (2009) Chem Biol 16:744
51. Cravatt BF, Demarest K, Patricelli MP, Bracey MH, Giang DK, Martin BR, Lichtman AH (2001) Proc Natl Acad Sci USA 98:9371
52. Cravatt BF, Giang DK, Mayfield SP, Boger DL, Lerner RA, Gilula NB (1996) Nature 384:83
53. Cravatt BF, Saghatelian A, Hawkins EG, Clement AB, Bracey MH, Lichtman AH (2004) Proc Natl Acad Sci USA 101:10821
54. Murphy RC, Fiedler J, Hevko J (2001) Chem Rev 101:479
55. Snyder F, Wood R (1969) Cancer Res 29:251
56. Dang L, White DW, Gross S, Bennett BD, Bittinger MA, Driggers EM, Fantin VR, Jang HG, Jin S, Keenan MC, Marks KM, Prins RM, Ward PS, Yen KE, Liau LM, Rabinowitz JD, Cantley LC, Thompson CB, Vander Heiden MG, Su SM (2009) Nature 462:739
57. Seltzer MJ, Bennett BD, Joshi AD, Gao P, Thomas AG, Ferraris DV, Tsukamoto T, Rojas CJ, Slusher BS, Rabinowitz JD, Dang CV, Riggins GJ (2010) Cancer Res 70:8981
58. Uetz P, Giot L, Cagney G, Mansfield TA, Judson RS, Knight JR, Lockshon D, Narayan V, Srinivasan M, Pochart P, Qureshi-Emili A, Li Y, Godwin B, Conover D, Kalbfleisch T, Vijayadamodar G, Yang M, Johnston M, Fields S, Rothberg JM (2000) Nature 403:623
59. Bulyk ML (2006) Curr Opin Biotechnol 17:422
60. Mangelsdorf DJ, Thummel C, Beato M, Herrlich P, Schutz G, Umesono K, Blumberg B, Kastner P, Mark M, Chambon P, Evans RM (1995) Cell 83:835

61. Bunting S, Gryglewski R, Moncada S, Vane JR (1976) Prostaglandins 12:897
62. Devane WA, Hanus L, Breuer A, Pertwee RG, Stevenson LA, Griffin G, Gibson D, Mandelbaum A, Etinger A, Mechoulam R (1992) Science 258:1946
63. Berge KE, Tian H, Graf GA, Yu L, Grishin NV, Schultz J, Kwiterovich P, Shan B, Barnes R, Hobbs HH (2000) Science 290:1771
64. Yang Q, Graham TE, Mody N, Preitner F, Peroni OD, Zabolotny JM, Kotani K, Quadro L, Kahn BB (2005) Nature 436:356
65. Schug TT, Berry DC, Shaw NS, Travis SN, Noy N (2007) Cell 129:723
66. Schug TT, Berry DC, Toshkov IA, Cheng L, Nikitin AY, Noy N (2008) Proc Natl Acad Sci USA 105:7546
67. Kinsey SG, Long JZ, O'Neal ST, Abdullah RA, Poklis JL, Boger DL, Cravatt BF, Lichtman AH (2009) J Pharmacol Exp Ther 330:902–910
68. Morozov VN, Morozova TY, Johnson KL, Naylor S (2003) Rapid Commun Mass Spectrom 17:2430
69. Rodriguez-Agudo D, Ren S, Hylemon PB, Redford K, Natarajan R, Del Castillo A, Gil G, Pandak WM (2005) J Lipid Res 46:1615
70. Kagaya S, Ohkura N, Tsukada T, Miyagawa M, Sugita Y, Tsujimoto G, Matsumoto K, Saito H, Hashida R (2005) Biol Pharm Bull 28:1603
71. Zhan Y, Du X, Chen H, Liu J, Zhao B, Huang D, Li G, Xu Q, Zhang M, Weimer BC, Chen D, Cheng Z, Zhang L, Li Q, Li S, Zheng Z, Song S, Huang Y, Ye Z, Su W, Lin SC, Shen Y, Wu Q (2008) Nat Chem Biol 4:548
72. Nicolaou KC, Evans RM, Roecker AJ, Hughes R, Downes M, Pfefferkorn JA (2003) Org Biomol Chem 1:908
73. Downes M, Verdecia MA, Roecker AJ, Hughes R, Hogenesch JB, Kast-Woelbern HR, Bowman ME, Ferrer JL, Anisfeld AM, Edwards PA, Rosenfeld JM, Alvarez JG, Noel JP, Nicolaou KC, Evans RM (2003) Mol cell 11:1079
74. Gavin AC, Bosche M, Krause R, Grandi P, Marzioch M, Bauer A, Schultz J, Rick JM, Michon AM, Cruciat CM, Remor M, Hofert C, Schelder M, Brajenovic M, Ruffner H, Merino A, Klein K, Hudak M, Dickson D, Rudi T, Gnau V, Bauch A, Bastuck S, Huhse B, Leutwein C, Heurtier MA, Copley RR, Edelmann A, Querfurth E, Rybin V, Drewes G, Raida M, Bouwmeester T, Bork P, Seraphin B, Kuster B, Neubauer G, Superti-Furga G (2002) Nature 415:141
75. Smith CA, O'Maille G, Want EJ, Qin C, Trauger SA, Brandon TR, Custodio DE, Abagyan R, Siuzdak G (2005) Ther Drug Monit 27:747
76. Wheeler DL, Barrett T, Benson DA, Bryant SH, Canese K, Chetvernin V, Church DM, DiCuccio M, Edgar R, Federhen S, Geer LY, Helmberg W, Kapustin Y, Kenton DL, Khovayko O, Lipman DJ, Madden TL, Maglott DR, Ostell J, Pruitt KD, Schuler GD, Schriml LM, Sequeira E, Sherry ST, Sirotkin K, Souvorov A, Starchenko G, Suzek TO, Tatusov R, Tatusova TA, Wagner L, Yaschenko E (2006) Nucleic Acids Res 34:D173
77. Sud M, Fahy E, Cotter D, Brown A, Dennis EA, Glass CK, Merrill AH Jr, Murphy RC, Raetz CR, Russell DW, Subramaniam S (2007) Nucleic Acids Res 35:D527
78. Vinayavekhin N, Saghatelian A (2009) ACS Chem Biol 4:617
79. Bucco RA, Zheng WL, Davis JT, Sierra-Rivera E, Osteen KG, Chaudhary AK, Ong DE (1997) Biochemistry 36:4009
80. Fiorella PD, Napoli JL (1991) J Biol Chem 266:16572
81. Kane CD, Bernlohr DA (1996) Anal Biochem 233:197
82. Smith CA, Want EJ, O'Maille G, Abagyan R, Siuzdak G (2006) Anal Chem 78:779
83. Baier LJ, Sacchettini JC, Knowler WC, Eads J, Paolisso G, Tataranni PA, Mochizuki H, Bennett PH, Bogardus C, Prochazka M (1995) J Clin Invest 95:1281
84. Levy E, Menard D, Delvin E, Stan S, Mitchell G, Lambert M, Ziv E, Feoli-Fonseca JC, Seidman E (2001) J Biol Chem 276:39679
85. Reitz J, Gehrig-Burger K, Strauss JF III, Gimpl G (2008) FEBS J 275:1790

86. Baker AR, McDonnell DP, Hughes M, Crisp TM, Mangelsdorf DJ, Haussler MR, Pike JW, Shine J, O'Malley BW (1988) Proc Natl Acad Sci USA 85:3294
87. Green S, Walter P, Kumar V, Krust A, Bornert JM, Argos P, Chambon P (1986) Nature 320:134
88. Spiegelman BM (1998) Diabetes 47:507
89. Zhang Z, Burch PE, Cooney AJ, Lanz RB, Pereira FA, Wu J, Gibbs RA, Weinstock G, Wheeler DA (2004) Genome Res 14:580
90. Noy N (2007) Biochemistry 46:13461
91. Tsukahara T, Tsukahara R, Fujiwara Y, Yue J, Cheng Y, Guo H, Bolen A, Zhang C, Balazs L, Re F, Du G, Frohman MA, Baker DL, Parrill AL, Uchiyama A, Kobayashi T, Murakami-Murofushi K, Tigyi G (2010) Mol Cell 39:421
92. Chakravarthy MV, Lodhi IJ, Yin L, Malapaka RR, Xu HE, Turk J, Semenkovich CF (2009) Cell 138:476
93. Krey G, Braissant O, L'Horset F, Kalkhoven E, Perroud M, Parker MG, Wahli W (1997) Mol Endocrinol 11:779
94. Chinetti G, Fruchart JC, Staels B (2000) Inflamm Res 49:497
95. Xu HE, Lambert MH, Montana VG, Parks DJ, Blanchard SG, Brown PJ, Sternbach DD, Lehmann JM, Wisely GB, Willson TM, Kliewer SA, Milburn MV (1999) Mol Cell 3:397

Index